W9-DGP-741

Good practice in science teaching

Good practice
in science teaching
What research
has to say

Edited by
Martin Monk and
Jonathan Osborne

Open University Press
Buckingham · Philadelphia

Open University Press
Celtic Court
22 Ballmoor
Buckingham
MK18 1XW

email: enquiries@openup.co.uk
world wide web: http://www.openup.co.uk

and
325 Chestnut Street
Philadelphia, PA 19106, USA

First Published 2000

Copyright © The editors and contributors 2000

All rights reserved. Except for the quotation of short passages for the purpose of
criticism and review, no part of this publication may be reproduced, stored in a retrieval
system, or transmitted, in any form or by any means, electronic, mechanical,
photocopying, recording or otherwise, without the prior written permission of the
publisher or a licence from the Copyright Licensing Agency Limited. Details of such
licences (for reprographic reproduction) may be obtained from the Copyright
Licensing Agency Ltd of 90 Tottenham Court Road, London, W1P 9HE.

ISBN   0 335 20391 4 (pb)   0 335 20392 2 (hb)

A catalogue record of this book is available from the British Library

Library of Congress Cataloging-in-Publication Data
Good practice in science teaching: what research has to say / edited
   by Martin Monk and Jonathan Osborne.
        p.   cm.
   Includes bibliographical references and index.
   ISBN 0-335-20392-2 (hb). – ISBN 0-335-20391-4 (pb)
   1. Science–Study and teaching.   I. Monk, Martin.
II. Osborne, Jonathan.
Q181.G5135   2000   99–29147
507'.1'2–dc21   CIP

Copy-edited and typeset by The Running Head Limited, www.therunninghead.com
Printed in Great Britain by Biddles Ltd, Guildford and King's Lynn

Contents

Contributors

All the contributors work in the School of Education, King's College London unless otherwise indicated.

Philip Adey is Professor of Science, Cognition and Education. He has worked as a chemistry teacher, an examiner, a science textbook author, a teacher trainer and researcher. He is the director of the Centre for the Advancement of Thinking, at King's College London.

Paul Black is an Emeritus Professor of Science Education. He has worked as a university lecturer, curriculum developer, materials author and educational researcher. He was the joint director of the Assessment of Performance Unit (science) with the late Rosalind Driver. He now continues to work to improve the use of assessment in education.

Margaret Cox is Professor of Information Technology in Education. She did her PhD in physics, directed the Computers in the Curriculum Project, and has authored programmes and researched their use, most notably as director of the ImPACT project. She is interested in all aspects of information technology.

Justin Dillon is a lecturer in science education. He has worked as a chemistry teacher in London and now lectures on the Master's course in Management. His research interests are the management of school science departments and environmental education; he undertakes much of his work through international collaboration.

Robert Fairbrother is a visiting lecturer at King's College London. He has worked as a physics teacher, examiner, curriculum developer, textbook author and teacher trainer. Bob's interests are in the fields of assessment and in helping children to learn.

Chris Harrison is a lecturer in biology education. She has worked as a biology teacher and headed a science department in a large London comprehensive. She is interested in cognitive acceleration and is researching and developing the use of formative assessment.

Sally Johnson is a lecturer in biology education. She has taught on vocational and A level programmes and works on in-service courses with teachers from overseas and as a science education consultant. Her research is into teacher development and change.

Carys Jones is a lecturer in applied language studies in education. She has worked as an EFL teacher overseas and has worked on science programmes for students whose mother tongue is not English. Her research interests are in literacy.

John Leach is a senior lecturer in science education at the University of Leeds. Prior to this he worked as a biology teacher and a researcher on the Children's Learning in Science (CLIS) project. He is interested in the development of conceptual understanding in science, epistemology and the public understanding of science.

Martin Monk is a lecturer in physics education and has taught physics in the UK and overseas. Currently he works on in-service courses for teachers from overseas and as a science education consultant. The focus of his research is classroom activities and teachers' styles.

Jonathan Osborne is a senior lecturer in science education. He has taught physics in London schools and was an advisory teacher in Inner London. He has undertaken research into young children's understanding of science, attitudes to science and science education for the public understanding of science.

Philip Scott is a lecturer in science education at the University of Leeds. He has worked as a physics teacher and is interested in both teaching and learning science concepts in high schools and the role of language in the acquisition of science concepts.

Shirley Simon is a lecturer in chemistry education. Before working as a research fellow she taught in London schools. Her research has explored primary teachers' ideas in science and progression in students' ideas. She is currently interested in argument in science.

Joan Solomon is a visiting professor at King's College and Professor of Science Education at the Open University. She taught physics in schools before working in teacher training at Oxford. She is one of the most prolific writers on science education in the UK.

Julian Swain is a lecturer in chemistry education. He taught in London schools before working as a research fellow on the Graded Assessment in Science Project. Currently he works on in-service education for overseas teachers and researches aspects of assessment.

Rod Watson is a senior lecturer in chemistry education. He has taught in schools in the UK and in Spain. The focus of his research is the role of pupil investigations and practical work in science education, and he has directed two major research projects in this area.

Introduction: research matters?

Jonathan Osborne and Martin Monk

Research in education is under attack. In the UK, Passmore (1999) reports the chief inspector of schools claiming, with little or no supporting evidence, that:

- teachers would learn how to teach better from their successful peers;
- teachers learn little from the 'wisdom of the professor of professional development';
- most educational research is 'irrelevant and impenetrable'.

More thoughtful concerns are raised in an ongoing debate (Hargreaves 1996; Hammersley 1997; Hargreaves 1997) that culminated with the Hillage report (Hillage *et al.* 1998). These latter authors argued that currently research is conducted by a fragmented research community which lacks coordination among their funders; that there was a lack of involvement of the teachers (the principal beneficiaries); and they raised concerns about the quality of some of the output. Furthermore, Hillage and co-authors felt dissemination was weak, with a lack of support and encouragement for academics to engage in the dissemination process. Drawing on research in science education, this book is an attempt to answer those criticisms: to argue that research has much to say that can inform the practice of science teaching and to disseminate research findings more widely.

The notion that teachers can learn *only* from each other is akin to arguing that doctors might discover a cure for malaria by watching each other's valiant attempts to treat the symptoms, rather than its cause. Essentially, education, like any profession, requires individuals to stand aside, to study both the minutiae of classroom practice and the broad sweep of both the policy and practice of its institutions, and to ask critical, reflective questions of what they see. Education requires questions whose answers often demand a level of specialist knowledge which is not accessible to those caught up in the relentless pressures of classroom life. For education is a multi-disciplinary profession, drawing as it does, on the more fundamental disciplines of psychology – to inform us about the nature of individuals and the learning process; philosophy – to inform us about the nature of the subject we teach and about the aims and values we espouse; history – as a treasure trove of case studies of how people have dealt with, and responded to similar issues in the past; and last, but not least, sociology – which informs us about the dynamics of the society in which we are situated and the values and concerns of the interested participants.

Standing at the crossroads of such disciplinary ideas, the task of the academic in education is to sift, to assimilate, and to distil the implications for the practice of science education. So, for instance, what are the messages from the evidence collected on the psychology of learning? What are the implications for pedagogy that follow from constructivist viewpoints? How does evidence on Piagetian-inspired interventions, painstakingly collected, re-orientate our ideas on sequencing and structuring activities for learning science? A knowledge of the philosophy of science might tell us about the shortcomings of the science we teach. Or, what does the work of linguists have to say about the nature of scientific language that makes science intrinsically difficult? These are all examples of the kind of questions that academic scholarship in education attempts to address; and are all examples of questions which practitioners find difficult to consider for more than a fleeting moment, when engaged in the mad rush of daily life generated often by the relentless drive to achieve improving standards.

Of necessity, research in science education is an investment in belief – belief that intellectual endeavour and focused study of particular aspects of learning and teaching will result in a better understanding of the predicaments faced by the learner and the teacher. Like any investments, there are winners and losers. But ultimately it is an act of faith that the products of such work will produce tangible improvements. Who could have foretold in 1976, that the work of Shayer and Wylam (1978) investigating the Piagetian levels of the schoolchildren, funded by the Social Sciences Research Council, would ultimately lead to the development of a course which has shown, and continues to show, significant improvements in exam results compared to other schools? Who could have foretold that the work of Rosalind Driver (1983), funded by the Secondary Science Curriculum Review, would lead to a major transformation in our understanding of the conceptual complexities of what it means to learn science? Who could have foretold that the sustained interest in assessment, its function and purpose shown by Paul Black and his colleagues over a period of 20 years, would lead to the invaluable insights on the role and function of assessment found in the recent publication *Inside the Black Box* (Black and Wiliam 1998)? Like any profession, there will be products which are mediocre; the PhD theses which languish in some dark and dusty corner; the journal articles that fail to reach the parts that others do. But, we would contend that this is the price that we have to pay, and *must* pay, if we are to acquire the few glittering prizes that provide evidence to take our knowledge and classroom practice forward a few faltering steps.

For, while the critics ardently point to the weaknesses, their alternative proposals have little merit. No other profession would attempt to dismantle the links to the research community that offer it new possibilities and improved practice. Doctors would be the first to support continuing medical research; engineers rely on research to provide them with novel materials and techniques; even lawyers, in that most maligned of professions, rely on legal analysts and academics for the advance of their practice. Why do so many casual commentators, outside the teaching profession, think that learning and teaching does not warrant research, just monitoring of performance? Is this the measure of the low regard that some politicians and members of the general public have for those who shape the future in our classrooms?

Teachers can certainly gain much personally from research that offers a valued opportunity for reflexive examination of their own practice. However, teachers' daily lives do not bring them into contact with the ever-growing body of contemporary theory that would help them analyse the nature of their problems, which ultimately will constrain their achievements. Moreover, there must also be a place for the large-scale academic research project which has the resources and contemporary expertise to address some of the many problems that exist within the teaching and learning of science. In short, the critics' remedies for the failings of educational practice are a recipe for stasis – the ossification of current practice as the epitome of what is best and the denial of hope of a better future.

In this book, we have drawn primarily on a body of expertise that resides in the science education research community at King's College London, asking colleagues to present, wherever possible, the understandings and implications that can be drawn from the wide body of research evidence that exists in their own specialist research interests in science education. Inevitably, in some areas, there is more, and in others, there is less. Perhaps not surprisingly, the 'jam' is unevenly spread but in all cases there has been no shortage. Rather, authors have presented the headlines – essential points for consideration within the limited confines of the space that we have allowed. In these chapters, there are points for consideration:

- within the daily round of classroom practice;
- within the department;
- for the wider issues of concern that recurrently surface in the teaching of science.

Inevitably, each chapter represents a partial view but we believe that, within the chapters of this book, there is much of substance that will both inform and challenge the practices of the reader. The evidence and scholarship that is presented is supported by detailed references at the end of each chapter. Works of particular importance have been asterisked as a first port of call for the interested reader who wishes to know more. While the complexity of educational research is such that the nature of the evidence rarely surpasses that seen in Yeats' 'the blue, the dim and the half-light', we hope the reader will find much here that will illuminate many aspects of their practice. In short, rather than science teaching being a practice which is the cumulation of years of ad hoc folk tales, we believe it is based on a well-established body of evidence that supports and justifies the practices of the classroom teacher. This book, then, is a contribution to setting the record straight, to arguing that the research community in science education does have much to say that would both inform and improve the practice of teachers, something which is recognized internationally if not nationally (Gibson 1994).

This book is also a tribute to the research and scholarship of the late Professor Rosalind Driver, whose work at The University of Leeds, and then at King's College London, was seminal in its inception. For she became one of the most pre-eminent figures in science education of her generation – a major figure on both national and international stages – who attracted considerable interest and respect from science education researchers and, most importantly, from science teachers. Throughout her professional career she displayed an enduring passion for science education and took very seriously the responsibility of research in trying to improve our understanding of what is involved in learning and teaching science and, indeed, what might constitute an education in science.

References

Black, P. and Wiliam, D. (1998) *Inside the Black Box: Raising Standards through Classroom Assessment*. London: King's College. (Also published, with minor changes and the same title, as an article in *Phi Delta Kappan*, 80(2): 139–48.)

Driver, R. (1983) *The Pupil as Scientist?* Milton Keynes: Open University Press.

Gibson, A. (1994) *International Collaborative Research in Education: A Study for the ESRC.* Swindon: Economic and Social Research Council.

Hammersley, M. (1997) Educational research and teaching: a response to David Hargreaves' TTA lecture, *British Educational Research Journal*, 23(2): 141–61.

Hargreaves, D.H. (1996) *Teaching as a Research-based Profession: Possibilities and Prospects.* Teacher Training Agency annual lecture 1996. London: TTA.

Hargreaves, D.H. (1997) In defence of research for evidence based teaching: a rejoinder to Martyn Hammersley, *British Educational Research Journal*, 23(4): 405–19.

Hillage, J., Pearson, R., Anderson, A. and Tamkin, P. (1998) *Excellence in Research on Schools* (Research Report RR74). London: Department for Education and Employment.

Passmore, B. (1999) Learn from each other, says Chief Inspector, *Times Educational Supplement*, 26 February.

Shayer, M. and Wylam, H. (1978) The distribution of Piagetian stages of thinking in British middle and secondary school children. II – 14- to 16-year-olds and sex differentials, *British Journal of Educational Psychology*, 48: 63–70.

Part I
The science classroom

When considering good practice in science teaching, as a priority our attention must turn to the student and the learning of science. Therefore, in considering what research has to say about good practice we can sensibly start with the learner as our primary focus. The first part of this book brings together seven chapters that do this by making the learner, and learning science, their principal concern.

The first three chapters look at research on how to help learners be more effective in their task. Then, the following three chapters (4, 5 and 6) look at learning science. Lastly, in this section, Chapter 7 looks at the affective: what students *feel* about learning science.

In Chapter 1 Robert Fairbrother has written on what we know about helping learners to become better learners by aiding them to think more carefully about what they are doing. In Chapter 2 Paul Black and Chris Harrison review work on the introduction and use of formative assessment. Both of these chapters are generic to the issue of being a learner and are applicable to learning in other subjects. John Leach and Philip Scott's chapter brings together issues raised by the first two under the umbrella of constructivism.

Constructivism, with a small *c*, has been around for a long time and the so-called Socratic method is recognizably a variety of constructivism. Just as Socrates used questions that enabled his students to build their knowledge by degrees, being careful always to start with what his students knew already, Leach and Scott articulate why this practice is so important in the learning of science.

Rod Watson's Chapter 4 turns our attention to what we know about students' learning through practical work and investigations. This chapter begs questions about the nature of science and, in particular, the relationship between evidence and theory. So, there follows Chapter 5, which looks at the nature of scientific knowledge. This chapter also looks at the research evidence on how much teachers know, and are influenced in their teaching by, a coherent philosophy of the nature of science. Carys Jones reminds us that to learn science is to learn to speak, read and write science. Her Chapter 6 looks at what we know about the signalling systems – pictures, diagrams, tables and graphs – as well as words – used in science and learning science. She shows how a more sensitive awareness to language issues by the teacher can improve students' learning.

The last section of this part of the book looks at students' feelings about science and learning science. Shirley Simon's Chapter 7 reviews evidence on what we know about the affective side to learning science, using this to show how science can be made more engaging for children.

To the casual reader this immediate focus on learning and the learner may appear to be slightly odd in a book, the title of which starts, 'Good practice in science *teaching*'. However, experienced teachers will often relate that the key to success is not to be found in the layman's notion of teaching as transmitting information from the head of the teacher to the head of the student. Similarly the message of research is that good practice in teaching builds on the idea that the learner must construct their own knowledge and skills for themselves through activities designed by the teacher. Rather than the naive simile of teaching being a process where the teacher is a full jug and the student an empty, passive and receptive beaker, research points to good practice in teaching as being one where the teacher is more of a tour guide and the student a traveller.

1 Strategies for learning

Robert Fairbrother

This chapter is about pupils regulating their own learning – an activity as important in science as in any other subject. Such self-regulation involves three main overlapping learning strategies:

- *cognitive* – the learning of the content of the subject (for example, electric charge, homeostasis), where pupils select relevant information from what is presented to them (notes copied from the board, the results of an experiment, and so on). They then have to elaborate and organize the information to add coherence, integrate new information with existing knowledge, and use various revision strategies;
- *metacognitive* – the *process* of learning. Pupils use a variety of planning, regulating and evaluating strategies and think about what they are doing when they are learning;
- *motivational* – which involves learning for its own sake, a belief in the value of the task and a belief in one's own abilities, i.e. self-efficacy.

Learning strategies are not necessarily the same as teaching strategies that teachers use, although there can be some overlap. The overlap occurs when good teaching strategies can be used by the pupils themselves; two examples are concept mapping, which is largely a cognitive strategy because it involves an understanding of scientific concepts and the relationships between them, and group discussions between pupils about understanding, which is largely a metacognitive strategy because it involves not just what pupils understand but how they gain their understanding.

There is also overlap between the three strategy areas themselves, and it is sometimes difficult to decide whether we are in the cognitive, metacognitive or motivational area. For example, a cognitive strategy which brings success can also be considered to be motivational. With these overlaps in mind, this chapter will emphasize the less well-known metacognitive and motivational strategies.

All learning strategies require that pupils be made explicitly aware that what they are doing can become a part of their own learning armoury. Pupils, and teachers, need to know a range of learning strategies because different strategies will suit different pupils. Variety in learning is as important as variety in teaching. More particularly, most learning strategies have to be taught to pupils since pupils need help in learning how to learn. Bergin (1996) found that high school pupils used very few learning strategies out of school, which indicates the need for more school instruction in learning strategies that encourages their transfer out of school. Much the same can also be said for teachers. In work done with some elementary school teachers using cooperative learning strategies, Nath *et al.* (1996) found that the teachers needed training, administrative support and peer encouragement in order to teach learning strategies. The above points can be summarized:

- teachers should be knowledgeable about different learning strategies;
- the teaching of strategies should take place on a regular basis;
- the teaching should follow a proper programme of development;
- the teaching of learning should be integrated with the teaching of the subject;
- pupils should be helped to adopt the strategies for their own independent use.

In order to cover as many issues as possible in a limited space, this chapter emphasizes conclusions and sometimes does not give the supporting evidence. Much of this is to be found in the references at the end of the chapter. (In particular, you can find more information about the issues mentioned above in Weinstein and Mayer 1986; Zimmerman and Martinez-Pons 1986, 1988; Pressley *et al.* 1989; Zimmerman 1989, 1994; Pintrich and De Groot 1990.)

Metacognition[1]

Being aware of the learning process, knowing 'what we know' and 'what we don't know' and thinking about how we learn is known as metacognition.

According to Dirkes (1985),the basic metacognitive strategies are:

- connecting new information to former knowledge;
- selecting thinking strategies deliberately;
- planning, monitoring, and evaluating thinking processes.

Metacognitive skills are needed when habitual responses are not successful. Guidance in recognizing, and practice in applying, metacognitive strategies will help pupils successfully solve problems throughout their lives. Studies show that increases in learning do follow direct instruction in metacognitive strategies. These results suggest that direct teaching of these thinking strategies can be useful, but that independent use develops gradually (Scruggs *et al.* 1985). Vermunt *et al.* (1995) has found that teaching thinking strategies and subject-specific knowledge together not only improved students' learning in the subject, but also showed transfer effects producing improved learning in another course.

However, while all pupils need help in learning metacognitive strategies, low attaining pupils need more help than do high attaining pupils. There is also some evidence (for example Loranger 1994) that unsuccessful students are less efficient in their use of learning strategies, and they perceive themselves as successful learners because they lack knowledge of their inefficient use.

Strategies for developing metacognitive behaviours

Identifying 'what you know' and 'what you don't know'

At the beginning of an activity pupils need to make conscious decisions about their knowledge. Initially you can get pupils to identify 'What I already know about . . .' and 'What I want to learn about . . .' They can learn to do this by looking at a 'pupil-speak' version of the National Curriculum or a summary of the topic to be taught. As the pupils find out more about the topic, they can learn to verify, clarify and expand, or replace with more accurate information, each of their initial statements. In this way there is a regular checking of what they are learning with what they want or are required to learn.

A method of teaching pupils how to identify what they know and what they don't know is for the teacher to divide a lesson into two halves. In the first half she starts by telling the pupils one or two learning outcomes which are expected. At the end of this half she asks the pupils whether the learning outcomes have been achieved and where in the lesson they occurred. She then starts the second half of the lesson by telling the pupils there will be two more learning outcomes but they have to identify these for themselves. Towards the end of the lesson she asks the pupils to say what they think the two learning outcomes were and provides an opportunity for everyone, including herself, to discuss them.

There is a wide variety of ways of adapting this strategy, such as asking individuals to write down what they think are the learning outcomes; asking pupils to discuss learning outcomes in pairs and to reach an agreement; giving the pupils a homework task which is to write down the learning outcomes and then in the next lesson asking them to discuss their decisions in groups. Yet another method is to ask pupils to identify what they *don't* know or are unsure about and to talk about it among themselves. There are two main requirements for this strategy to work: allow time for the strategy to be discussed and practised in the lesson; and show the pupils how they can use the strategy at any time at home on their own or with a friend.

Teachers in the LeAP (Learning Advancement for Pupils)[2] project who tried this strategy found that it worked well but that they had to be careful not to do it too often. A period of self-reflection every lesson tended to be counter-productive.

Talking about thinking

Talking about thinking is important because pupils need a thinking vocabulary so that they are able to communicate with each other and with the teacher. During planning and problem-solving situations, teachers should think aloud so that pupils can follow demonstrated thinking processes. Giving examples in this way and encouraging discussion helps to develop the vocabulary that pupils need for thinking and talking about their own thinking. Clarifying and labelling thinking processes when pupils use them is also important for pupil recognition of thinking skills.

Group work, properly organized, is particularly useful for developing communication skills and self-reflection. Grimes (1995) has found that low-attaining students prefer group learning and are weaker on cognitive, goal-orientated and effort-related learning strategies. Meece and Jones (1996) analysed self-reports of some fifth- and sixth-grade pupils, and found that the pupils reported greater confidence and mastery motivation in small groups than in whole classes, and greater work avoidance in whole-class than in small-group lessons.

One group strategy is paired problem-solving in which one pupil talks through a problem, describing his thinking processes. His partner listens and asks questions to help clarify thinking. Similarly, in reciprocal teaching (Palinscar et al. 1986), small groups of pupils take turns at playing teacher, asking questions, and clarifying and summarizing the material being studied.

Talking about one's understanding is difficult. Pupils have to develop a vocabulary in order to express themselves, they also need the confidence to say what they know and, particularly, what they don't know. Talking to each other in small groups is a non-threatening situation which also breaks away from the conventional lesson in which the teacher controls the questions and often – the answers. Pupils are often so accustomed to having too little time to compose an answer or to think of a question that they do not even try or assume that they are faced with a rhetorical mannerism of the teacher (such as the continued repetition of 'OK?' which, despite the interrogatory way in which it is used, is not really intended as a question requiring an answer). They thus have to become accustomed to being in control of the situation and to having time to think (see the work of Budd-Rowe 1974, on 'wait time'). Even people who have a good grasp of the subject often require several seconds in order to collect their thoughts and frame an answer or compose a question. It is much more difficult for pupils who are meeting something for the first time to do this. Furthermore, they have to develop social skills and skills of listening so that they can contribute positively to any discussion. Towns and Grant (1997) found that university students who engaged in a regular Friday afternoon cooperative discussion session, which focused on conceptual issues rather than on ways of solving problems, not only moved away from rote learning strategies and towards strategies which enabled them to integrate concepts over the whole semester, but also developed interpersonal skills and communication skills.

Another example is reported by Yu and Stokes (1998). They organized a 'teaching studio' in which small groups of students discussed problems given in a lecture and prepared answers to be given to the whole class. The session was divided into clear activities, for example, a short lecture, discussion or problem-solving (in small groups), or presentation of solutions. None of the students in each group knew beforehand who would be chosen by the lecturer to make the presentation, and the representative eventually chosen had to make the presentation without notes. This motivated all the students to try to understand the issues involved. Students in a group were given marks based on the quality of the presentation of their representative. This encouraged cooperation in the group to ensure everyone understood and could give a clear explanation. A variation of this is described by Eunsook Kim in Fairbrother *et al.* (1998).

> As a preparation, I divided the content to be covered into several steps. Then questions were made for each step so the students could follow the content by answering the questions. [There were five questions covering the steps to be taken in calculating a magnetic field.]
>
> I did one example (an infinitely long straight wire) on the board, and then I asked the students to do a second example themselves. I asked them to talk to their neighbours, and I also said I would ask them questions afterwards. After a few minutes, I named five students to answer the five questions. I gave them a few minutes to prepare their answer and then I asked the first student to answer question 1, the second student question 2, and so on. [The process was repeated for several examples each, requiring the calculation of the magnetic field produced in different ways, for example, a toroid, an infinitely long solenoid].

Many of the above strategies can be seen as teaching strategies. However, they are capable of being turned into learning strategies if the process the pupils are going through is made explicit to them so that they can use it themselves, and the value of the strategy is made clear.

Keeping a thinking journal

Another means of developing metacognition is through the use of a journal or learning log. This is a diary in which pupils reflect upon their thinking, make notes of their awareness of ambiguities and inconsistencies, and comment on how they have dealt with difficulties. This journal is a diary of progress. An example of this is reported by Lan (1996) in a study in which the performance of three groups of students was compared. One group, the self-monitoring group, recorded the frequency and intensity of their various learning activities. A second group, the instructor-monitoring group, evaluated the instructor's teaching. A third group was a control group which did the same course without any special treatment. The self-monitoring group performed better in course tests, used more self-regulated learning strategies and developed better knowledge representation of the course content.

The status of such pupil's journal is important. It belongs to the pupil who might, or might not, wish to share it with a teacher or a colleague. It is essential, however, to help the pupils to improve their journal and hence their thinking. One way of doing this is for teachers to learn from their own diary attempts and to pass on the messages to their pupils. Teachers can also produce an extract from their 'own' journal or from that of a volunteer pupil, discuss it with the whole class and then provide an opportunity for pupils to talk to each other about the extract and about their own journals.

Planning and self-regulation

Pupils must assume increasing responsibility for planning and regulating their learning. It is difficult for learners to become self-directed when learning is planned and monitored by someone else. Most teaching and learning situations are planned and directed by teachers. Pupils realize this and usually wait to be told what to do. If they are required to make decisions for themselves without being told that the initiative has been passed to them and without having been educated in decision-making, the lesson usually descends into chaos. This was experienced by many teachers when investigations in science were introduced into the National Curriculum. It is now generally recognized that extended project work, properly planned and managed, encourages the development of learning strategies which are essential for life-long learning.

Pupils can be taught to make plans for learning activities, including estimating time requirements, organizing materials, and scheduling procedures necessary to complete an activity. Such plans fall into an area which many people call study skills (see the section on motivation which says a little more about this). Criteria for the evaluation of learning must be developed with pupils so they learn to think and ask questions of themselves as they proceed through a learning activity.

In an earlier publication, Fairbrother (1995), I have described some successes and failures in working with pupils and in helping them to assess their own understanding and progress. To help them with their revision, pupils were provided with summaries of the work just done. To help them make progress in doing investigations, the pupils were given specially written versions of the national curriculum. These ideas have been tried by, and indeed were copied from, other teachers. The failures occurred when the pupils were not given sufficient guidance about how to work on their own. The successes came when the pupils were shown how to check their work against the curriculum and were given time to talk about it among themselves.

One of the difficulties in helping pupils to plan and regulate their learning is to find the right level of detail with which to describe to the pupils what they are expected to know, understand and do. In attempting to achieve complete clarity, one can easily be led down the route to a multitude of specific behavioural objectives and to an atomistic view of learning. The inverse of this leads to a few generalized statements which pupils cannot understand and use. There is a good discussion of this problem and references for further reading, in the context of mastery learning, in Postlethwaite and Haggarty (1998).

Debriefing the thinking process

Closure and summarizing activities focuses pupil discussion on thinking processes to develop an awareness of strategies that can be applied to other learning situations.

A three-step method is useful.

- First, the teacher guides the pupils in a review of the activity. This involves clarifying the thinking processes which the pupils have used and their feelings about what they have done and learned.
- Then, the group classifies related ideas, identifying thinking strategies used; for example, preparation before the activity started, monitoring during the activity and revision when it was completed.
- Finally, they evaluate their success, discarding inappropriate strategies, identifying those valuable for future use, and seeking promising alternative approaches.

White and Fredriksen (1998) have found that a process of reflective assessment in which pupils reflect on their own and other's inquiries has a particular benefit for low-achieving pupils.

Self-evaluation

Guided self-evaluation experiences can be introduced through individual discussions and checklists focusing on thinking processes. Gradually, self-evaluation will be applied more independently. As pupils recognize that learning activities in different disciplines are similar, they will begin to transfer learning strategies to new situations (see Vermunt 1995).[3]

Concept maps, organized hierarchically, can be used in several of the above strategies both by teachers when teaching, and by pupils to help their own learning. Hierarchical concept maps (see, for example, Novak and Gowin 1984; Novak 1990; Sizmur 1996) help readers clarify ambiguities of text and reveal misconceptions. They also enable the learner to initiate ideas that can be shared visually with someone else. For self-evaluation, both during and after the teaching of a unit of work, concept maps allow pupils to keep track of their learning and to update their understanding.

Establishing the metacognitive environment

A metacognitive environment encourages awareness of thinking. While individual teachers and departments can do something on their own, ideally it is a whole-school activity in which planning is shared between teachers, school library media specialists, and pupils, where thinking strategies are discussed and evaluation is ongoing. Schraw (1998) describes three instructional strategies for promoting the construction and acquisition of metacognitive awareness: promoting general awareness, improving self-knowledge and regulatory skills, and promoting learning environments that are conducive to the construction and use of metacognition.

A key requirement is to find the time to tell pupils about different learning strategies, to give them opportunities to practise the strategies and to exchange ideas and experiences about what they have learned and how they did the learning. There is some indication, however, that the introduction of a national curriculum encourages the use of surface learning strategies rather than deep strategies. Hacker and Rowe (1997), in a study of the impact of a national curriculum in Australia, found that teachers placed more emphasis on lower-order intellectual skills. There were also fewer speculative behaviours and fewer behaviours concerned with experimentation. Many of the teachers suggested that the reason for this was an overburdened curriculum.

This situation is exacerbated when, as in England and Wales, it is accompanied by the publication of test results particularly from tests whose content is constrained by the limitations of space and time so that there is an emphasis on factual questions which can be answered quickly. In science, for example, there is little coverage of such things as data analysis, debate about conflicting evidence and drawing conclusions, all of which require extended answers. In this situation Fairbrother *et al.* (1995) found that teachers had an increased tendency to teach to the test and so to emphasize the recall of knowledge. Other evidence, for example, Birenbaum and Feldman (1998), indicates that pupils are less anxious and more inclined to use learning processes when they know that they will be assessed through open-ended questions rather than through multiple-choice questions (MCQ). In a study with older students, Scouller (1998) obtained similar results.

Learning skills and study skills

In the literature there is reference to learning skills and study skills. The distinction between these, on the one hand, and learning strategies on the other, is not clear. At the extremes a learning *skill* is a surface strategy which is specific to a particular task and tends to concentrate on rote learning, while a learning *strategy* is a deep strategy which is transferable across different tasks and different subjects and leads to deeper understanding. An example of a learning skill is the repetition of small bits of text, which are then strung together when learning a passage of text. Another is constant repetition until an association is established. Many pupils have been taught their multiplication tables by the whole class chanting them together until an association is established – in the minds of many people 'six sevens' triggers off the response '42'. Yet another is making up mnemonics to learn a series of facts (for example ROYGBIV for the colours of the rainbow). The term 'study skills' tends to be used for such things as management of one's time for revision, dividing a subject into sections for ease of study, and working through past examination questions.

School policies and teaching practices

In an attempt to give pupils success, many teachers shield pupils, usually low-ability or disadvantaged pupils, from the effort which learning requires. It is important not to allow pupils to evade difficult academic tasks and not to diminish the need to make an effort. To do so is not only to be condescending to pupils, something which they usually perceive and which damages their self-esteem, but also to store up trouble for the future when difficult tasks cannot be avoided and effort is needed to solve them. Where the problem of boys is concerned, provided the right task is chosen, anecdotal evidence indicates that the presentation of it as 'tough' and 'challenging' – words which they see as having male connotations – encourages them to tackle it with more enthusiasm.

Having high expectations, giving encouragement and rewarding success are important. Yet, again, this must be done carefully. Research shows that failing pupils interpret actions such as sympathy, praise for modest achievement, or help given when it is not asked for as signs that they lack ability.

Research findings are sometimes rather contradictory about some of the ways of helping pupils to make an effort to learn. Some teachers have found that making expected learning outcomes clear helps pupils to identify what is important in a lesson and so improves learning. Others say that giving this assistance deprives the pupils of working out the ideas for themselves and of directing their own learning. Without some assistance, however, learning can head in the wrong direction. Some research done during the graded assessment movement in the 1980s found that often the teacher of a lesson had one set of objectives, the pupils questioned after the lesson had identified a different set, and the researcher who sat in on the lesson had a third set.

The answer lies in finding the right balance – between keeping pupils in the dark and telling them everything. One way of doing this was suggested by Brown (1987), and makes use of cognitive strategies, particularly of organization and elaboration, and domain-specific knowledge (Alexander and Judy 1988; Alexander *et al.* 1994). While aimed at improving the way lectures are given in a university, it is applicable to teaching in schools and has the additional advantage that it can be used by pupils to help them make better notes.

As a result of research with students, Brown suggested four structuring moves for improving lectures:

- *signposts* – these indicate the structure and direction of the lesson:
  'Today I want to deal briefly with . . .'
  'First, I will provide a definition of . . .'
  'Secondly, outline a model of . . .'
  'Thirdly, point out some limitations . . .'
- *frames* – these are statements which indicate beginnings and endings of subtopics and topics:
  'So, that's the essential outline of . . .'
  'Let's have a look now at how . . . is used'
  Without well-defined frames of topics and sub-topics, a student's notes can become a mish-mash.
- *foci* – these are statements and emphases which highlight key points:
  'Now this step is very important'
  'So the main point is this (pause) . . .'[4]
- *links* – these are phrases and statements which link one part of the lecture to another, and the whole lecture to the experience of the students.

It is important to emphasize that, if we expect pupils to use these structuring moves for themselves, attention must be drawn to the moves as they are used and pupils must be helped to use them themselves by giving examples. It is helpful to adapt the strategy described earlier and to ask pupils periodically to write down the two (or three?) most important points that have been made in the past 15 minutes. A comparison with what other pupils have written or with what the teacher had in mind is instructive and revealing. If the teacher has made good use of frames and foci, there should be good agreement between the teacher and his class.

Motivation[5]

Pupils will learn only if they want to learn. This is the single most important factor in raising standards, but the educational reforms of the past three decades (from Nuffield to the National Curriculum) have concentrated more on changing the organization and structure of the educational system and on attempting to improve the quality of teaching and teaching materials. There are many aspects of motivation but four seem to be particularly important:

- *the learning culture* – this is mainly about peer pressure and the ethos which exists in the school;

- *intrinsic purpose of learning* – this shifts the emphasis from extrinsic motives for learning such as getting better examination grades and marks to intrinsic motives such as a belief in learning for its own sake and a mastery of the subject;
- *self-efficacy* – this concerns pupils' belief in themselves and their ability to learn;
- *task value* – this is about the pupils' interest in the task and their perceptions of its importance and value.

The learning culture

Evidence from the US (US Department of Education 1992) shows that pupils themselves believe their own ability and effort, rather than the quality of their teachers or of their textbooks, are the main reasons for school achievement. Peer pressure and the culture of the school have a particularly strong effect on how they use their ability and how hard they work.

- They want to be seen to be able rather than hard-working because they believe that hard-working pupils risk being considered either excessively ambitious or of limited ability, both of which they find embarrassing. Also, if they are lazy, the remedy lies in their hands whereas, if they are 'dim', the remedy is out of their control.
- To avoid unpopular labels, pupils – especially the brightest – believe they must strike a balance between the extremes of achievement, not too high and not too low.
- Many pupils adopt an attitude of indifference to hard work, a stance that implies both confidence in their own ability and a casual regard for academic success.
- At the extreme, many low-achieving pupils deny the importance of learning and withhold the effort it requires in order to avoid the stigma of having tried and failed. As Holton (1995:73) says, 'Many, if not most, of the pupils entering the school have experienced failure to such a degree that they have developed an empirical formula of "no work" so that it cannot be criticised.' Delaying tactics in lessons – such things as creating a disturbance, not bringing a pen, not opening their bags and not getting out their books – are all attempts to avoid having to do work which will expose their inability.

Peer pressure is particularly strong in the teens. There is a conflict between wanting to stay on at school to get good examination grades and a reluctance to be seen to be ambitious and hard-working. It might be that the low attainers encourage this reluctance and even spend their time actively developing it in order to hide their own lack of success. In this way a culture is developed which rejects academic aspirations. This culture might be confined to particular groups in a school, particularly when pupils with this attitude are grouped together and encourage each other, but it can also extend to the whole school when such pupils are spread around different classes in an attempt to dilute the effect. Peer pressure exists, not just in classrooms, but also, and perhaps more so, in the social interactions which occur outside classes.

One particular sub-group or sub-culture is that of boys. Pheonix (1998) has found that most young, teenage boys believe that working hard at school and being seen to be clever are not male attributes. She says that, to be properly masculine, boys think they have to be good at sport and to be tough and confrontational. Inside this outer shell of toughness there is, however, an uncertainty which causes them constantly to try to justify themselves in relation to the above notion of maleness. Those who are clever or who work hard maintain their maleness by decrying their achievements and they retain their standing in the eyes of their peers by being cheeky to teachers.

The solution to these problems is easy to say but difficult to put into practice: use peer pressure to develop a culture which supports high expectations. In the case of boys, for example, we can make use of the uncertainty which Pheonix says boys have of their masculine identity by changing their view. One way of doing this is to build an image of masculinity which includes hard work and academic achievement by identifying and emphasizing role models who have these characteristics.

Intrinsic purpose of learning

While some research (for example Ames 1992) indicates that learning for its own sake is related to better use of learning strategies and to higher achievement, moving motivation in this direction is particularly difficult. The solution for teachers is a matter of stating the obvious – make the subject as interesting as possible by using a variety of teaching and learning strategies – and ensure that these strategies lead to success. Somewhat paradoxically one might establish intrinsic purposes by making use of the extrinsic pressures of classwork and homework marks. If, through these pressures, pupils see they are achieving some success, it can lead them to a greater interest in the subject and hence to learning it for its own sake.

For many pupils an impending test is a big motivating factor, albeit extrinsic. Tuckman (1996) compared an incentive motivation strategy (a weekly test) with a learning strategy (identifying key terms in a textbook, writing definitions of them and elaborating on them) and found that the incentive strategy group did better than the learning strategy group. However, the effect was greater for low-attaining students than for high- and medium-attaining students. What he did not do was to find the effect of using both strategies. The work of Schutz and Lanehurt (1994) seems to indicate that a mixture of incentive strategies and learning strategies is needed.

One way of mixing these strategies is to ask pupils to think how they will be asked to show their knowledge and understanding in a test. This leads them to think about the kinds of questions which will be asked. If they then try to write their own questions, exchange them with a colleague, answer the questions, discuss the answers and compare the questions with those in past test papers, they will not only be better prepared for the forthcoming test but will also acquire a useful learning strategy.

Some specific advice is often given to pupils under the heading of study skills or learning skills, particularly in the revision period leading up to an examination. Leaving such advice until the last minute is, however, a mistake. The suggestions should be offered and practised as soon as pupils are capable of reflecting upon what they are doing. The main parts of the advice are:

- *clarify why you are learning* – this can be anything from 'because I am interested' to 'if I don't, I will get low marks'. The advice given to students is to try to see that learning has positive advantages such as gaining a qualification which will lead to a greater choice of career. Schutz and Lanehart (1994) have found that, when long-term educational goals are supported by the achievement of educational subgoals, and the use of a sufficient number of useful learning strategies, academic performance tends to improve.
- *establish targets* – these can range from relatively broad and long-term goals to more immediate and more directly motivational objectives such as, 'At the weekend I will review the module on Electricity and Magnetism and will be clear about the main learning outcomes.' The advantage of short-term, achieveable, objectives is that they give quick rewards – 'Jam today, not tomorrow';[6]
- *establish priorities* – be clear about deadlines and match work to meet them. Partition time so that all necessary tasks are covered – this includes time for leisure but particularly time for doing the most uninteresting tasks such as revising the most boring subject;[7]
- *give rewards* – link rewards to short-term targets, for example have a cup of coffee when a task is finished. Build variety into your work. 'When I have done the bit I don't like, I'll go on to the bit I do like.';
- *review* – regularly review what has been achieved and what has not been achieved. Keep a record of progress.

Self-efficacy

As might be expected, students' belief in themselves is more of a problem with low achievers than with high achievers. High achievers also seem to make more use of cognitive and metacognitive learning strategies.

The greater a pupil's self-efficacy, the more effort the pupil is likely to make in order to learn. However, the relationship between effort and attainment is complex. Most educators believe that all pupils should learn as much as their ability and effort will permit. Yet, most schools, parents, and all league tables of performance recognize high attainment alone. High-ability pupils are usually rewarded with the highest grades, marks and publicity, while the efforts of less-able pupils are seldom acknowledged, and their attainments do not inspire effort. Hence, low-ability pupils and those who are disadvantaged – pupils who must work hardest – have the least incentive to do so. They find this relationship between high effort and low attainment unacceptable, something to be avoided if possible. Some of them express their displeasure by simple indifference, others by disruption and deception.

Task value

The following is a paraphrase of a comment made by an American high school student:

> I leave for school at 8 o'clock. I have hockey practice or a music lesson until about 5.30. I get home at about 6.30 and I want something to eat and then see my friends. On Saturdays I go to work. What more do you want?

In England the answer from most teachers would be at least an hour of homework each night, and the student might say, 'What sort of life is that?'

Striking the right balance between work and play is difficult and depends on what is seen as most valuable. Many parents say they would rather their children be happy, well behaved and have good friends than get good examination results. Employers say they look for good social skills such as the ability to cooperate and work in a team. Schools say they aim to produce good, well-rounded citizens able to take their part in society. The main emphasis in schools, however, is in getting good examination results. All this gives pupils conflicting messages about what is important.

If academic achievement is the top priority, and most teachers believe that this is the case and that everything else will follow, we must make this completely clear to pupils. It is wrong to have this as the priority for able pupils but to have a different priority for low-ability pupils.

Conclusions

This chapter has explored a number of factors that affect a pupil's ability to respond and learn from the experience provided by teaching. It has sought to show that explicit attention to the cognitive, metacognitive and motivational aspects of learning can not only assist pupil's learning, but also provide them with a set of skills that transcend the science classroom and learning science – skills that will hopefully be of benefit to lifelong learning.

Notes

1 The structure of this section on metacognition is adapted from an ERIC digest, ED 327218, *Developing Metacognition*, by Elaine Blakey and Sheila Spence, which

is a reprint of an article by the same authors 'Thinking for the future' in *Emergency Librarian*, 17(5) May–June 1990: 11–14. It summarizes some of the outcomes of recent research into metacognition. More information can be obtained from the ERIC Clearinghouse on Information & Technology, Syracuse University, eric@ericir.syr.edu Any additional material which is not specifically acknowledged comes from the Learning Advancement for Pupils (LeAP) Project at King's College London, University of London.

2 A research and development project based at King's College London, 1996–98.

3 There is a useful discussion in van der Stoep *et al.* 1996 about the similarities and differences between natural sciences, social sciences and humanities. It covers the views which teachers have of their subject, the ways the subjects are assessed and the influences of these on the motivation of students and the learning strategies they use.

4 Often foci are neglected by lecturers, yet they are crucial for note-taking and learning. Occasionally lecturers insert foci *after* important points. They may use remarks such as 'What I have been talking about is very important'. The glazed look on students' faces on hearing this suggests that backward focusing is not advisable.

5 The structure of this section is based upon US Department of Education (June 1992), *Hard Work and High Expectations: Motivating Students to Learn*. Office of Educational Research and Improvement, Programs for the Improvement of Practice.

6 Only pupils who are experienced in self-regulation will be able to do this on their own. Targets might be too ambitious or not ambitious enough. They should, therefore, be flexible and be capable of being updated. Most pupils will need the help of a teacher. It is also helpful to discuss targets with a friend, a colleague or a member of the family. Irrespective of how it is done, the satisfaction of having achieved a target is very rewarding.

7 Most pupils and, indeed, many adults are disorganized. They do not know how to keep an appointments/work diary and they do not know how to estimate and allocate time for different tasks. Pupils need help in doing these things, and there is a great deal of comfort to be gained from having a clear view of the way ahead. Many schools ensure pupils keep a homework diary, but this needs to be developed into a more active and useful tool.

References

Alexander, P. and Judy, J. (1988) The interaction of domain-specific and strategic knowledge in academic performance, *Review of Educational Research*, 58: 374–404.

Alexander, P., Kulikowich, J. and Jetton, T. (1994) The role of subject-matter knowledge and interest in the processing of linear and non-linear text, *Review of Educational Research*, 64: 201–52.

Ames, C. (1992) Classrooms: goals, structures and student motivation, *Journal of Educational Psychology*, 84: 261–71.

Bergin, D.A. (1996) Adolescents' out-of-school learning strategies, *Journal of Experimental Education*, 64(4): 309–23.

Birenbaum, M. and Feldman, R.A. (1998) Relationships between learning patterns and attitudes towards two assessment formats, *Educational Research*, 40(1): 90–8.

*Blakey, E. and Spence, S. (1990) Thinking for the future, *Emergency Librarian*, 17(5): 11–14.

Brown, G. (1987) Lores and laws of lecturing, *Physics Bulletin*, 38: 305–7.

*Budd-Rowe, M. (1974) Relation of wait-time and rewards to the development of language, logic and fate control: part II – rewards, *Journal of Research in Science Teaching*, 11(4): 291–308.

Dirkes, M.A. (1985) Metacognition: students in charge of their thinking, *Roeper Review*, 8(2): 96–100.

*Fairbrother, R. (1995) Pupils as learners, in R. Fairbrother, P. Black and P. Gill (eds) *Teachers Assessing Pupils: Lessons from Science Classrooms*. Hatfield: Association for Science Education.

Fairbrother, R.W., Dillon, J.S. and Gill, P. (1995) Teacher assessment at Key Stage 3: teachers' attitudes and practices, *British Journal of Curriculum and Assessment*, 5(3): 25–31, 46.

Fairbrother, R., Jae-hyeok Choi, Sung-Won Hwang, Sun-Shin Jung, Duckun Kim, Eunsook Kim, Hyoung-Suck Kim, Jung-Won Lee, Jong-Ah Soh and Jin Yoon (1998) Different teaching strategies – some experiences in Korea, unpublished paper. Seoul National University, Korea.

Grimes, S.K. (1995) Targeting academic programs to student diversity utilizing learning styles learning-study strategies, *Journal of College Student Development*, 36(4): 422–30.

Hacker, R.G. and Rowe, M.J. (1997) The impact of national curriculum development on teaching and learning behaviours, *International Journal of Science Education*, 19(9): 997–1004.

Holton, J. (1995) Pupil assessment without stress: a study of assessment in an EBD school, in R. Fairbrother, P. Black and P. Gill (eds) *Teachers Assessing Pupils: Lessons from Science Classrooms*. Hatfield: Association for Science Education.

Lan, W.Y. (1996) The effects of self-monitoring on students' course performance, use of learning strategies, attitude, self-judgement ability and knowledge representation, *Journal of Experimental Education*, 64(2): 101–15.

Loranger, A.L. (1994) The study strategies of successful and unsuccessful high school students, *Journal of Reading Behavior*, 4: 347–60.

Meece, J.L. and Jones, M.G. (1996) Gender differences in motivation strategy use in science: are girls rote learners? *Journal of Research in Science Teaching*, 33(4): 393–406.

Nath, L.R., Ross, S. and Smith, L. (1996) A case study of implementing a cooperative learning program in an inner-city school, *Journal of Experimental Education*, 64(2): 117–36.

Novak, J.D. (1990) Concept maps and vee diagrams: two metacognitive tools to facilitate meaningful learning, *Instructional Science*, 19(1): 29–52.

Novak, J. and Gowin, D. (1984) *Learning How to Learn*. New York: Cambridge University Press.

Palinscar, A.S., Ogle, D.S., Jones, B.F., Carr, E.G. and Ransom, K. (1986) *Teaching Reading as Thinking*. Alexandria, VA: Association for Supervision and Curriculum Development.

Pheonix, A. (1998) Paper presented to the British Association for the Advancement of Science, *Daily Telegraph*, 10 September.

*Pintrich, P.R. and De Groot, E. (1990) Motivational and self-regulated learning components of classroom academic performance, *Journal of Educational Psychology*, 82: 33–40.

*Postlethwaite, K. and Haggarty, L. (1998) Towards effective and transferable learning in secondary school: the development of an approach based on mastery learning, *British Educational Research Journal*, 24(3): 333–53.

*Pressley, M., Borkowski, J. and Schneider, W. (1989) Good information processing: what it is and what education can do to promote it, *International Journal of Educational Research*, 13: 857–67.

*Schraw, G. (1998) Promoting general metacognitive awareness, *Instructional Science*, 26: 113–25.

Schutz, P.A. and Lanehart, S.L. (1994) Long-term educational goals, subgoals, learning strategies use and the academic performance of college students, *Learning and Individual Differences*, 6(4): 399–412.

Scouller, K. (1998) The influence of assessment method on students' learning approaches: multiple choice question examination versus assignment essay, *Higher Education*, 35(4): 453–72.

Scruggs, T.E., Mastropieri, M.A., Monson, J. and Jorgenson, C. (1985) Maximizing what gifted pupils can learn: recent findings of learning strategy research, *Gifted Child Quarterly*, 29(4): 181–5.

Sizmur, R.S. (1996) Collaborative concept mapping and children's learning in primary science. Unpublished PhD thesis, School of Education, King's College London.

Towns, M.H. and Grant, E.R. (1997) 'I believe I will go out of this class knowing something': co-operative learning activities in physical chemistry, *Journal of Research in Science Teaching*, 34(8): 819–35.

Tuckman, B.W. (1996) The relative effectiveness of incentive motivation and prescribed learning strategy in improving college students' course performance, *Journal of Experimental Education*, 64(3): 197–210.

*US Department of Education (1992) *Hard Work and High Expectations: Motivating Students to Learn*. Washington, DC: Office of Educational Research and Improvement, Programs for the Improvement of Practice.

van der Stoep, S.W., Pintrich, P.R. and Fagerlin, A. (1996) Disciplinary differences in self-regulated learning in college students, *Contemporary Educational Psychology*, 21: 345–62

Vermunt, J.D. (1995) Process-oriented instruction in learning and thinking strategies, *European Journal of Psychology of Education*, 10(4): 325–49.

*Weinstein, C.E. and Mayer, R. (1986) The teaching of learning strategies, in M. Wittrock (ed.) *Handbook of Research on Teaching*, 315–27. New York: Macmillan.

White, B.Y. and Fredriksen, J.R. (1998) Inquiry, modeling and metacognition: making science accessible to all students, *Cognition and Instruction*, 16(1): 3–118.

Yu, K.N. and Stokes, M.J. (1998) Students teaching students in a teaching studio, *Physics Education*, 33(5): 282–5.

*Zimmerman, B.J. (1989) A social cognitive view of self-regulated academic learning, *Journal of Educational Psychology*, 81: 329–39.

Zimmerman, B.J. (1994) Dimensions of academic self-regulation: a conceptual framework for education, in D.H. Schunk and B.J. Zimmerman (eds) *Self-regulation of Learning and Performance: Issues and Educational Applications*. Hillsdale, NJ: Lawrence Erlbaum Associates.

Zimmerman, B.J. and Martinez-Pons, M. (1986) Development of a structured interview for assessing student use of self-regulated learning strategies, *American Educational Research Journal*, 23: 614–28.

Zimmerman, B.J. and Martinez-Pons, M. (1988) Construct validation of a strategy model of student self-regulated learning, *Journal of Educational Psychology*, 80: 284–90.

Internet references

ERIC Clearinghouse on Information & Technology Syracuse University Center for Science and Technology, 4th Floor, Room 194 Syracuse, New York 13244–4100
eric@ericir.syr.edu
ERIC Digests
gopher://ericer.syr.edu:70/11/Digests
ERIC digests are in the public domain and may be freely reproduced and disseminated.
United States Department of Education
http://www.ed.gov

2 Formative assessment

Paul Black and Chris Harrison

As well as planning schemes of work and teaching lessons, teachers need to monitor students' learning. Formative assessment is the obvious means of providing students with feedback to improve their performance. The term 'formative assessment' does not have a tightly defined and widely accepted meaning. In this chapter it is interpreted as encompassing all those activities, undertaken by teachers, and/or by their students, which provide information to be used as feedback to modify the teaching and learning activities in which they are engaged.

This chapter is based on a recent review of the research evidence on formative assessment (Black and Wiliam 1998a). The first of the three main sections will discuss some aspects of the research evidence that indicate guidelines for improving classroom practice. The second discusses research evidence about the quality of such practice. The third section then discusses the implications for practice which follow from the evidence.

Features of research on formative assessment

Evidence of success

From the literature published since 1986, it is possible to select at least 20 studies which describe how the effects of formative assessment have been tested by quantitative experimental-control comparisons. All of these studies show that innovations which strengthen the practice of formative assessment produce significant learning gains. These studies range over all ages (from 5-year-olds to university undergraduates), across several school subjects, and over several countries. The experimental outcomes are reported in terms of effect size, which is the ratio of the net mean gain in learning score to the standard deviation of the pupils' scores. Typical effect sizes for the experiments reviewed are between 0.4 and 0.7: an effect size of 0.4 would mean that the average pupil involved in an innovation would record the same achievement as a pupil in the top 35 per cent of those not involved, which would correspond to a gain of between one and two grades at GCSE.

Several of these studies exhibit another important feature: they show that improved formative assessment helps the 'low achievers', and also pupils with learning disabilities, more than the rest, and so reduces the spread of attainment while also raising it overall. Any gains for such pupils could be particularly important, for any 'tail' of low educational achievement can signify wasted talent. Furthermore, pupils who come to see themselves as unable to learn usually cease to take school seriously – and resort to disruption or to truancy.

The research reports also bring out other features, namely:

- There is a need to enhance feedback between those taught and the teacher, thereby calling for *significant changes in classroom practice*.
- For assessment to function formatively, the results have to be used to adjust teaching and learning, so indicating a need *to make teaching programmes more flexible and responsive*.
- Formative assessment, like all other ways to generate effective learning, requires that *pupils be actively involved*.
- Attention should be given to the ways in which assessment can affect *the motivation* and *self-esteem of pupils*, and to the benefits of *engaging pupils* in self- and peer assessment.

The learning gap

In a general analysis of feedback, Ramaprasad (1983: 4) states that 'Feedback is information about the gap between the actual level and the reference level of a system parameter which is used to alter the gap in some way.'

He argued that, for feedback to exist, the information about the gap must be used to alter the gap, otherwise there is no effective feedback. Sadler (1989), writing more specifically about school learning, emphasized that action will be inhibited if the gap, between the state revealed by feedback (i.e. what the student knows and understands at the start of learning) and the desired state or 'reference level' (i.e. the knowledge and understanding which is the target for the learning), is seen as impracticably wide. He further argued that ultimately, the action to close that gap must be taken by the student – a student who automatically follows the diagnostic prescription of a teacher without understanding of its purpose or orientation will not learn. Thus self-assessment by the student is essential.

To develop this notion further, it is helpful to look more closely at the two elements – the goal of the learning, and the action taken to appraise and to close the gap.

Goals of learning

Within published research, there are examples to show that, in many cases, pupils lack any clear overview of their learning and, that where self-assessment is involved, pupils are easily confused because they do not understand the criteria on which to base such assessment. In addition, many of the successful innovations have developed self- and peer assessment by pupils as ways of enhancing formative assessment. The main problem that those developing self-assessment encounter is not the problem of trustworthiness – pupils are generally honest and reliable in assessing both themselves and one another – rather it is that pupils can only assess themselves when they have a sufficiently clear picture of the targets that their learning is meant to attain. Surprisingly, and sadly, many pupils do not have such a picture, and appear to have become accustomed to receiving classroom teaching as an arbitrary sequence of exercises with no overarching rationale. Claxton has argued that this is a particularly serious problem in science education, describing the difficulty in striking terms:

> Secondary science on the other hand was like being on a train in carriages with blanked out windows. You were going in a single direction, about which you had no choice. The train stopped at every station and you had to get off whether you liked it or were interested or not . . . because the windows were opaque you could not see the countryside in-between, so you did not know how the stations were linked or related to each other.
>
> (Claxton 1991: 25)

It requires hard and sustained work to overcome this culture of passive reception. The examples in Box 2.1 illustrate the ways in which such work can be rewarding. They also show how some science teachers found that, for formative assessment to be productive, pupils should be trained in self-assessment.

This line of argument shows that pupils can only play an effective part in their own assessment within programmes designed to help them achieve, and sustain, an overview of their learning targets, and then to apply the relevant criteria to their own progress. Such a programme should aim to translate curriculum aims into language that all pupils can understand, and at a level of detail that helps them relate directly to their learning efforts. This

---

Box 2.1   Examples of how formative assessment was productive

A group of secondary science teachers met at King's College between 1991 and 1993 to share ideas about developing formative assessment in their classrooms. A book describing the outcomes includes chapters written by eight of them (Fairbrother *et al.* 1995). The following extracts illustrate some of their findings.

Here, two teachers comment on their scheme to develop pupils' self-assessment by asking each pupil to fill in, on a personal record sheet, boxes, opposite the items in a list of attainment statements, to show how well they had achieved them.

> Most pupils are honest in their own assessment for most of the time. Some, especially the less able, don't like to admit that they are not coping and sometimes say they understand when they do not. Teachers try to help by emphasizing to each pupil that the record is a private document between the two of them and that what matters is that the teacher can see where the pupil has problems so that help can be given where needed . . . Many at the start write very few, and very vague, comments, but during the year these change and become more explicit and perceptive and so more useful.
>
> (Fairbrother *et al.* 1995: 15–16)

Another teacher used 'Pupil Assessment and Revision Sheets' listing learning aims (e.g. 'I can describe how a spring stretches and explain why it can be used in a newtonmeter') on which each pupil had to assess their learning in detail. Year 7 pupils found this difficult, but there was clear success with Year 10:

> They have told us that they are now able to focus their learning on to the areas which they feel they have the least confidence with. The staff have noticed that when they are revising a unit with a class, the students can pinpoint which parts/concepts in the unit they have the most difficulty with. Instead of revising everything in the unit sketchily, the teacher can home in on the real problems.
>
> (Fairbrother *et al.*1995: 32)

In a third school similar pupil sheets were used to alleviate and change pupils' concerns about tests. Evidence of success in the first year of the innovation gave promise of long-term advantages:

> We have often found that pupils are worried about impending tests and express horror at the thought of them . . . However, when questioned about their fears, it appears that the major causes of consternation result from uncertainty; not knowing what will be required of them . . . [As a result of the innovations pupils] will gain self knowledge and place less emphasis on the written tests thus removing the natural fears that these often engender and giving opportunities for pupils to consider their results on a more objective basis both amongst themselves and with teachers and parents.
>
> (Fairbrother *et al.* 1995: 46–9)

issue is brought out in more direct terms by a science teacher in New Zealand:

> Think of it from the kid's point of view, the kid gathers information from what you've given them already, they filter it, decide what's relevant to them, they interpret what they need to do however they like, they act on that information, and then from whatever you do, or from whatever things happen, they gather more information and so on.
>
> (Bell and Cowie 1997: 27)

The orientation by a student toward his or her work can only be productive if that student shares the teacher's notions of what constitutes quality in the subject. However, there is no subject for which the criteria of quality can be reduced to a simple set of unambiguous rules. Those who have tried to do this have usually found it necessary to accompany rules with several exemplars (anyone who has tried to standardize the work of a team of examiners will be familiar with this necessity). Thus the student has to learn as much by examples as by precept. It has been found that giving students the work of others to mark, or requiring pupils to make up problems rather than to answer them, produces larger learning gains than the conventional exercises in which pupils tackle problems set by others (see Rosenshine *et al.* 1996).

The directions of desirable change are to be chosen in the light of the learning aims of science education. Thus both pedagogy and assessment have to be fashioned in the light of assumptions, both about how learning is best achieved and about what it means to learn science, which will help to construct a model of progress in learning a topic in physics. Any teaching plan must be founded on such a model, which should guide the construction of formative assessment procedures and of items of helpful diagnostic quality, and could set criteria for grading summative assessments. This idea lay behind the 10- (now 8-) level system for the National Curriculum in England and Wales (Black 1997, 1998).

Bridging the gap

The goals set for the learner are to be chosen by the teacher acting in the role of supporter rather than as director of learning. A key concept here was introduced by Wood *et al.* (1976) and expressed by the metaphor of 'scaffolding' – the teacher provides the scaffold for the building, but the building itself can only be constructed by the learner. In this supportive role, the teacher has to set targets which are both attainable in the short term and adequately modest in relation to the learners' prospects of success.

However, it is only by grasping and deploying an understanding of the targets that the learner can come to evaluate his or her own performance. The student's appraisal of performance will be important, and fruitful, in so far as the task is appropriate to the goals, and in so far as it has been tackled in the light of some understanding of them. There will clearly be an iteration between any attempt at that task, an evaluation of that attempt in the light of the aims, and a clarification of the aims to guide further attempts.

Brown and Ferrara (1985) highlight one facet of low achievement, by showing that low-achieving students lack self-regulation skills, and that when these are taught to them, they progress rapidly. Low achievement may also arise because of a breakdown in communication – a gap in mutual understanding between the learner and those who possess the expertise being taught – rather than because of some defect inside the learner. Thus a student might try to revise work on problems posed by a set of electrical circuit 'puzzle boxes' by learning by heart the solutions for each example, whereas the purpose of the work was to help the student grasp the concepts of current and voltage.

Self-assessment at the point of learning is a crucial component for developing complex understandings through reflective habits of mind. Within a framework of formative assessment as an integral part of learning, it is an essential development, as well as a potentially powerful source for the improvement of learning. Indeed, some have argued that metacognition, by which they mean awareness and self-direction about the nature of their learning work, is essential to students' development in concept learning, and the work described here is clearly serving that purpose (Brown 1987; White and Gunstone 1989). Thus improved formative assessment can lead to changes which should be a powerful help in helping students to become more effective learners.

Much of what has been said above can be applied to peer assessment as well as to self-assessment. There have been extensive studies of peer collaborative learning (Wood and O'Malley 1996) and several of these bear out that peer assessment can be both reliable and productive. Sadler (1989) argues that, in peer assessment, pupils will be evaluating work at the same level as their own; that each can see a wide range of examples of performance; that other pupils' work will provide a range of examples of faults for critical discussion; and that, in such evaluation, pupils will be less inhibited by defensiveness than they might be with their own work. However, peer groups have to be working with formative aims and should not see the work as an example merely of grading one another. The example presented in Box 2.2 shows how a physics unit was used as the basis of an experiment to test the value for learning of developing self- and peer assessment: the brief account given serves to illustrate both the style of some of the relevant studies and some of the typical results obtained.

Pupils' expectations and self-esteem

The involvement of pupils in their own assessment changes both the role of the pupil as learner and the nature of the relationship between teacher and pupil, making the latter shoulder more of the responsibility for learning and calling for a radical shift in pupils' own perspectives about learning. A science teacher in Spain describes a difficulty in achieving this aim:

Box 2.2   An example taken from physics

White and Frederiksen (1998) describe an innovative teaching course for middle school pupils which emphasized practical inquiry for learning about force and motion. The work involved 12 classes of 30 pupils each, in two schools, all taught to a carefully constructed curriculum plan in which conceptually based issues were explored through experiments and computer simulations. The experimental work was structured around the pupils' use of tools of systematic and reasoned inquiry. There was emphasis on communication skills, and all of the work was carried out in peer groups. Each class was divided into two halves for one fraction only of their class time: in this fraction, a control group worked on a general discussion about improving the module, while an experimental group used this same time for discussion of their performance, structured to promote reflection through both peer and self-assessment.

All pupils were given the same basic skills test at the outset. The outcome measures included a mean score on projects throughout the course and a score on a test of the physics concepts involved. On the project scores, the experimental group showed a significantly better overall score than the control; however, when the pupils were divided into three groups according to low, medium or high scores on the initial basic skills test, the low scoring group showed a superiority, over their control group peers, of more than three standard deviations, the medium group just over two, and the high group just over one. A similar pattern of gains was also found for the concepts test.

This science project worked with a version of formative assessment, which was an intrinsic component of a more thorough-going innovation to change teaching and learning. The experimental-control difference here lay not in extra assessment work by teachers, but in the focus on self- and peer assessment achieved by giving pupils opportunities to reflect on their learning. Two other distinctive features of this study were first, the use of outcome measures of different types, but all directly reflecting the aims of the teaching, and secondly, the differential gains between pupils who would have been labelled 'low ability' and 'high ability', respectively.

The idea of self-evaluation is a difficult one, because the students don't fully comprehend the idea and only really think in terms of their exam mark. Generally speaking they don't reflect on their own learning process in an overall fashion. [They think] their assessment has more to do with the effort they made than with what they have actually learnt. In fact the main reason why students fail is their lack of study techniques, since they still tend to try to simply memorise things.

(Black and Atkin 1996: 99)

The expectations that pupils have built up from their experiences of assessment in school can actually constitute an obstacle to their taking a positive role in assessment. This is supported in an account of a study of primary pupils in the Geneva Canton (Perrin 1991). Here, it emerged that the pupils believed that summative assessments were for the school's and their parents' benefit, not for themselves. Since the assessment was not used to tell them how to work differently, they saw it as a source of pressure, which made them anxious. As a consequence of such evidence, the Canton decided to reduce its summative tests and enhance the formative role of assessment.

Ironically, where formative assessment has been emphasized, it has been found that pupils bring to the work a fear of assessment from their experience of summative tests. They share with teachers a difficulty in converting from norm-referenced to criterion-referenced ways of thinking. This matters because, as long as pupils compare themselves with others, those with high attainment are too little challenged and those with low attainment are demotivated. Feedback to any pupil should be about the particular qualities of the work, with advice on what the pupil can do to improve. It should avoid comparisons with other pupils. A research study which illustrates the importance of these issues is described in Box 2.3.

The ultimate user of assessment information, which is elicited in order to improve learning, is the pupil. Here there are two aspects – one negative, one positive. Where the classroom culture focuses on rewards – gold stars, grades or place-in-the-class ranking – then pupils look for the ways to obtain the best marks rather than at the needs of their learning, which these marks ought to reflect. In consequence, pupils try to avoid difficult tasks and spend time and energy looking for clues to the *right* answer. Many are reluctant to ask questions out of fear of failure. Pupils who achieve poor results come to believe that they lack ability and to attribute their difficulties to a defect in themselves which they cannot repair. So they *retire hurt*, avoid investing effort in learning, which could only lead to disappointment, and try to build up their self-esteem in other ways. While the high achievers can do well in such a culture, the overall result is to enhance the frequency and the extent of under-achievement of the majority.

Thus the effectiveness of formative feedback depends upon several detailed features of its quality, and not on its mere existence or absence. It seems also that there can be differential effects, between low and high achievers, for any type of feedback.

The quality of practice

Many research studies show that the everyday practice of assessment in classrooms is beset with problems and shortcomings, as the following quotations, taken from such studies, illustrate:

Box 2.3   The importance of feedback on performance improvement

Butler (1988) set out to explore a theory about a link between intrinsic motivation and the type of evaluation that pupils have been taught to expect. The experiment involved 48 11-year-old Israeli pupils selected from 12 classes across four schools. They were taught material not directly related to their normal curriculum, and were required to tackle written tasks individually under supervision, with an oral introduction and explanation. Each pupil received one of three types of written feedback with returned work. One-third of the classes were given individually composed comments on the match, or not, of their work with the criteria which had been explained to all of them beforehand. A second group were given only grades, derived from the scores on their session's work. The third group were given both grades and comments. Scores on the work done in, and for, the teaching sessions served as outcome measures.

For the *comments only* group, their scores increased by about one-third between the first and second sessions, for both types of task, and remained at this higher level for the third session. For the *grades only* group, their scores declined on both tasks between the first and last sessions. The *comments with grade* group showed a similar decline in scores. This last feature seems to indicate that, even if feedback comments are, in principle, operationally helpful for a pupil's work, their effect can be undermined by the negative motivational effects of normative feedback, i.e. by giving grades.

Tests of pupils' interest also showed a similar pattern: however, the only significant difference between high- and low-achieving students was that interest was undermined for the low achievers by either of the regimes involving feedback of grades, but was not affected where the feedback was by comments alone. High achievers in all three feedback groups, however, maintained a high level of interest.

The results are consistent with research in other literature which indicates that *ego-involving* feedback, i.e. feedback which encourages an introverted concern with one's own self-image rather than on what one should do to reach the target, is less effective than *task-involving* feedback. Even the giving of praise can lower rather than enhance the work of low achievers, and pre-occupation with grade attainment can lower the quality of task performance (Butler and Neuman 1995).

Assessment remains the weakest aspect of teaching in most subjects. Despite improvement, it remains poor overall in almost one school in eight in Key Stage 3 . . . Although the quality of formative assessment has improved perceptibly, it continues to be a weakness in many schools.

(Ofsted 1998: 88 and 91/2)

Why is the extent and nature of formative assessment in science so impoverished?

(Daws and Singh 1996: 99)

Marking is usually conscientious but often fails to offer guidance on how work can be improved. In a significant minority of cases, marking reinforces under-achievement and under-expectation by being too generous or unfocused. Information about pupil performance received by the teacher is insufficiently used to inform subsequent work.

(Ofsted 1996: 40)

The detailed points that are established in these and related studies (including several in countries outside the UK) may be summarized in three sets of points. Inevitably, these are summaries and do not apply to every classroom. The first set is concerned with *effective learning*:

- Teachers' tests encourage rote and superficial learning; this is seen even where teachers say they want to develop understanding.
- The questions and other methods used for assessment are not shared between teachers in the same school, and are not audited in relation to what they actually assess.
- For primary teachers particularly, there is a tendency to emphasize quantity and presentation of work and to neglect its quality in relation to learning.

The second set is concerned with *negative impact* of assessments:

- The giving of marks and the grading functions are over-emphasized, while the giving of useful advice and its learning function are under-emphasized.
- It is common for teachers to use approaches which appear to pupils to emphasize competition rather than personal improvement. In consequence, assessment feedback teaches pupils with low attainments that they are not able to learn, so they are demotivated.

The third set focuses on *the managerial role* of assessments:

- Teachers' feedback to pupils often seems to serve social and managerial functions, at the expense of the learning functions.
- Teachers are often able to predict pupils' results on external tests – because their own tests imitate them – but at the same time they know too little about their pupils' learning needs.
- The collection of marks to fill up records is given greater priority than the analysis of pupils' work to discern learning needs.

Implications for practice

The evolution of effective teaching

The research studies show that effective programmes of formative assessment involve far more than the addition of a few observations and tests to an existing programme. Indeed, it is clear that instruction and formative assessment are indivisible. But, how can teachers improve their own assessment work? The following sections give several guidelines based on research.

Choose tasks carefully – plan for formative assessment.

The choice and structure of tasks to stimulate learning is of central importance. Tasks have to be justified in terms of the learning aims that they serve, and they can only work well if opportunities for pupils to communicate their evolving understanding are built into the planning. This point is underlined by the research study quoted in Box 2.4.

Use a wide variety of sources and types of evidence.

Many teachers already do this, as was found in a study of ten science teachers in New Zealand:

> The sources of formative assessment for the teachers included the teachers' observations of the students working, for example, in practical activities; the teachers reading student written work in their books, posters, charts and notes; and the teachers listening to students' speech, including their existing ideas, questions and concerns, and the new understandings they were developing. The teachers set up different learning situations to provide opportunities for this information to be gathered or elicited. For example, the teachers organised practical and investigative work, brainstorming, spot tests, students recording their before-views, library projects, watching a video, whole class discussions and student self-assessment activities.
>
> (Bell and Cowie 1997: 23)

In all such work, the quality of the task or questions, i.e. their relevance to the main learning aims, and their clear communication to the pupil, needs scrutiny: teachers should collaborate with each other, and draw – critically – on outside sources, to collect such tasks and questions.

Bell and Cowie also found it useful to distinguish between 'planned formative assessment' and 'interactive formative assessment', the latter being part of any classroom dialogue. Some of the guidelines such as choice of tasks clearly relate to the planned aspects. Others such as thinking time for questions relate to the interactive aspect, while others again involve both.

Box 2.4   An example of paying attention to learning aims

The work of Watson *et al.* (1998) with 32 science teachers explored their difficulties with science investigations (AT 1). They found that, where investigations were left very open, pupils did not receive useful feedback. They describe how giving structure helped with this:

> The lessons were structured so that the pupils were given many opportunities for evaluation of investigative procedures and provided with opportunities to improve their investigations in response to comments and criticisms. For example, after deciding which independent variable they would investigate, the pupils planned how they would make their measurements. This was followed by a class discussion to allow groups to tell their plans to the rest of the class and receive any comments. The discussion was guided by the teacher who asked questions such as 'How will you measure that?' 'How will you make it more accurate?'. After further discussion in groups, each pupil filled in a structured planning sheet supplied by the teacher. The planning sheet was collected by the teacher and used to guide her discussion with the pupils in the next lesson.
>
> (Watson *et al.* 1998: 3)

A different finding is described in this next extract from their paper:

> Another significant problem was that there was a mismatch between teachers' aims for investigations and the things pupils consider they had learnt during the investigations. Teachers' aims usually feature both procedural and conceptual aspects, while pupils' descriptions of what they had learnt was heavily biased towards the conceptual aspects, linked to the questions being investigated.
>
> (Watson *et al.* 1998: 4)

Questions should be for thinking – so give pupils time to think about their responses.

Questions from the teacher are a natural and direct way of checking on learning, but are often unproductive. What often happens is that a teacher answers her or his own question after only two or three seconds. Where a minute (say) of silent thought is not tolerable, there is no possibility that a pupil can think out what to say. There are then two consequences. One is that, because the only questions that can produce answers in such a short time are questions of fact, these predominate. The other is that pupils don't even try to think out a response – if you know that the answer, or another question, will come along in a few seconds, there is no point in trying. The teacher then lowers the level of questions and so keeps the lesson going while being out of touch with the understanding of most of the class – the question–answer dialogue becomes a ritual and thoughtful involvement suffers. These difficulties are found in many classrooms, as the following quotation shows:

> The quality of teachers' questioning is very variable in the degree to which it extends pupils' thinking, draws out their ideas, and encourages them to volunteer points and explore further, thus providing evidence of achievement. Too often, teachers engage in closed questioning, limiting pupils' responses or even neglecting to take up issues that pupils raise, and ultimately failing to register how far they have understood the objectives of the work.
>
> (Ofsted 1998: 92/3)

There are several possible ways to evoke thoughtful responses and so break this particular cycle:

- Ask pupils to discuss their thinking in pairs, or in small groups, so that any respondent is speaking on behalf of others.
- Use a multiple choice format for a question and ask pupils, after being given time to think, to vote on the options, and then ask some to justify their vote.
- Ask all to write down an answer and then ask a selected few to read out what they have written.

What is essential in any approach is that pupils be given time to think, and that any dialogue should evoke thoughtful reflection in which all pupils can be encouraged to take part. Only then can the formative process start to work.

Prepare dialogue that provokes pupils' ideas and take all their ideas seriously.

Question and answer sequences are only one aspect of classroom dialogue. Discussions, in which pupils are led to talk about their understanding in their own ways, provide the opportunity for the teacher to respond to, and re-orientate the pupil's thinking. However, teachers often respond, quite unconsciously, in ways that inhibit the learning of a pupil. Recordings commonly show that the teacher is looking for a particular response and, lacking the flexibility or the confidence to deal with the unexpected, tries to direct the pupil towards giving the expected answer (Filer 1995; Pryor and Torrance 1996). Over time, pupils get the message – they are required not to think out their own answers but to guess at the answer that the teacher expects.

Feedback should be linked to criteria of learning, not to norms or marks or grades.

The research studies quoted above have shown that, if pupils are given only marks or grades, they do not benefit from the feedback on their work. Feedback has been shown to improve learning where it gives each pupil specific guidance on strengths and weaknesses. Thus, the way in which test results are reported back to pupils is a critical feature: the feedback to pupils should not be an overall mark, but an identification of their own strengths and weaknesses, together with the means and opportunities to work with this evidence to overcome difficulties. While it may also be necessary to have a summative test with marks at the end of the teaching of a block or module, such a test cannot be of much value for formative purposes.

For pupils to be independent learners – convey aims, promote self- and peer assessment.

The first step here is to help pupils to understand the aims of each piece of work and to share the teacher's criteria for quality. Then they can attempt self-assessment, and so can produce, as part of that assessment, valuable evidence about the nature and causes of any difficulties. Peer assessment can also be valuable because, in appraising one another's work, pupils can learn what the criteria mean by trying to apply them and by justifying their findings to others.

Learn from your pupils – and to respond to what you learn.

To pitch feedback at the right level to stimulate and help is a delicate skill. A teacher's approach should start by confronting the question, 'Do I really know enough about the understanding of my pupils to be able to help each of them?' This is a formidable target, but the challenge to practise is to give more priority to meeting it as fully as possible by learning from pupils' own products and responding to the needs that these indicate.

Believe that all your pupils can learn.

The beliefs that teachers hold about the potential of all of their pupils to learn make a difference to how they respond to them. To sharpen the contrast by overstating it, there is, on the one hand, the *'fixed IQ'* view, implying that each pupil has a fixed, inherited, intelligence, so that one must accept that some can learn quickly and others can hardly learn at all. On the other hand, there is the *'untapped potential'* view, prevalent in some non-western cultures, which starts from the assumption that so-called 'ability' is a complex of skills that can be learned. Here, the underlying belief is that all pupils can learn more effectively if one can clear away the obstacles set up by previous difficulties, be they cognitive failures that have never been diagnosed, or damage to personal confidence, or a combination of the two. Clearly, the truth lies between these two extremes, but the research evidence is that ways of managing formative assessment, which work with the assumptions of 'untapped potential', do help all pupils to learn; furthermore, they can give particular help to those who have previously fallen behind.

Obstacles to success

All these points make clear that improving formative assessment is a difficult enterprise. Some pupils will resist attempts to change, for change is threatening, and an emphasis on the challenge to think for yourself (and not just work harder) can be disturbing. Pupils will find it hard to believe in the benefits of changes before they have experienced those benefits.

Most teachers will be concerned because many of the relevant initiatives take more class time, at least initially, as new procedures are developed. When a central purpose is to change the outlook on learning and the working methods of pupils, all involved will need time to adapt. Thus, teachers have to take risks in the belief that such investment of time will yield rewards in the future. However, the alternative – 'delivery' and 'coverage' with poor understanding – is often pointless and may even be harmful.

The improvement of formative assessment cannot be a simple matter: there is no 'quick fix'. On the contrary, the rewards promised by the research evidence can only be secured by a teacher who finds his or her own ways of incorporating the ideas that are set out above into classroom work. Such changes can only happen slowly, through sustained programmes of professional development and support (Black and Wiliam 1998b). This does not weaken the message here – indeed, it should be a sign of its authenticity, for lasting and fundamental improvements in teaching and learning can only happen in this way.

References

Bell, B. and Cowie, B. (1997) *Formative Assessment and Science Education: Summary Report of the Learning in Science Project (Assessment)*. Hamilton, NZ: Centre for Science, Mathematics and Technology Education Research, University of Waikato.

Black, P.J. (1997) Whatever happened to TGAT? in C. Cullingford (ed.) *Assessment vs. Evaluation*. London: Cassell.

Black, P. (1998) *Testing: Friend or Foe? Theory and Practice of Assessment and Testing*. London: Falmer Press.

Black, P.J. and Atkin, J.M. (1996) *Changing the Subject: Innovations in Science, Mathematics and Technology Education*. London: Routledge for OECD.

*Black, P. and Wiliam, D. (1998a) Assessment and classroom learning, *Assessment in Education*, 5(1) 7–71.

*Black, P. and Wiliam, D. (1998b) *Inside the Black Box: Raising Standards through Classroom Assessment*. London: King's College. (Also published, with minor changes and the same title, as an article in *Phi Delta Kappan*, 80(2): 139–48.)

Brown, A. (1987) Metacognition, executive control, self-regulation and other mysterious mechanisms, in F.E. Weinert and R.H. Kluwe (eds) *Metacognition, Motivation, and Understanding*. Hillsdale, NJ: Lawrence Erlbaum.

Brown, A.L. and Ferrara, R.A. (1985) Diagnosing zones of proximal development in culture, communication and cognition, in J.V. Wersch (ed.) *Vygotskian Perspectives*. Cambridge: Cambridge University Press.

*Butler, R. (1988) Enhancing and undermining intrinsic motivation; the effects of task-involving and ego-involving evaluation on interest and performance, *British Journal of Educational Psychology*, 58: 1–14.

Butler, R. and Neuman, O. (1995) Effects of task and ego-achievement goals on help-seeking behaviours and attitudes, *Journal of Educational Psychology*, 87(2): 261–71.

Claxton, G. (1991) *Educating the Enquiring Mind: The Challenge for School Science*. New York: Harvester Wheatsheaf.

Daws, N. and Singh, B. (1996) Formative assessment: to what extent is its potential to enhance pupils' science being realised? *School Science Review*, 77(281): 93–100.

*Fairbrother, B., Black, P.J. and Gill, P. (eds) (1995) *Teachers Assessing Pupils: Lessons from Science Classrooms*. Hatfield: Association for Science Education.

Filer, A. (1995) Teacher assessment: social process and social product, *Assessment in Education*, 2(1): 23–38.

Ofsted (1996) *Subjects and Standards. Issues for School Development Arising from Ofsted Inspection Findings 1994–5. Key Stages 3 and 4 and post-16*. London: HMSO.

Ofsted (1998) *Secondary Education 1993–7: A Review of Secondary Schools in England*. London: HMSO.

Perrin, M. (1991) Summative evaluation and pupil motivation, in P. Weston (ed.) *Assessment of Pupils Achievement: Motivation and School Success*. Amsterdam: Swets and Zeitlinger.

Pryor, J. and Torrance, H. (1996) Teacher–pupil interaction in formative assessment: assessing the work or protecting the child? *The Curriculum Journal*, 7(2): 205–26.

Ramaprasad, A. (1983) On the definition of feedback, *Behavioral Science*, 28: 4–13.

Rosenshine, B., Meister, C. and Chapman, S. (1996) Teaching students to generate questions; a review of the intervention studies, *Review of Educational Research*, 66(2): 181–221.

*Sadler, R. (1989) Formative assessment and the design of instructional systems, *Instructional Science*, 18: 119–44.

*Watson, J.R., Goldsworthy, A. and Wood-Robinson, V. (1998) One hundred and twenty hours of practical science investigations: a report of teachers' work with pupils aged 7 to 14. Conference paper, Copenhagen Royal School of Education, Practical Work in Science Education.

White, B.Y. and Frederiksen, J.R. (1998) Inquiry, modeling and metacognition, *Cognition and Instruction*, 16(1): 3–118.

White, R.T. and Gunstone, R.F. (1989) Meta-learning and conceptual change, *International Journal of Science Education*, 11: 577–86.

Wood, D. and O'Malley, C. (1996) Collaborative learning between peers: an overview, *Educational Psychology in Practice*, 11(4): 4–9.

Wood, D., Bruner, J.S. and Ross, G. (1976) The role of tutoring in problem solving, *Journal of Child Psychology and Psychiatry*, 17: 89–100.

3   Children's thinking, learning, teaching and constructivism

John Leach
and Philip Scott

In this chapter, we will present what we see as being the most important messages for practice from research on learning and teaching science. First, we consider what is known about how children learn science. Attention is then focused on the implications of these insights for planning and implementing science teaching.

What do we know about how children learn science?

In order to address the issue of how children learn science, we will first consider the characteristics of the scientific knowledge to which they are being introduced.

Some characteristics of the scientific knowledge introduced in the school curriculum

If asked to describe the nature of scientific knowledge, many lay adults might say that it is a precise description of how some aspect or other of the natural world functions. If asked where scientific knowledge comes from, they might say that it is based on careful observations produced by great scientists experimenting in their laboratories. Of course, such accounts of the nature of science are very limited. Indeed, most professional scientists, historians, philosophers and sociologists of science would probably argue that this kind of account *misrepresents* the nature of science in fundamental ways. For example, scientific concepts such as gravity are much more than descriptions of objects falling to the ground. The scientific notion of gravitational field is not 'there to be seen' by anyone, great scientist or otherwise. It is inappropriate, therefore, to characterize the concept of gravitational field as a description of some aspect of the world. Although the mythology of science tells the story of Isaac Newton 'discovering' gravity when an apple fell onto his head, this misrepresents the ways in which individual scientists interact with one another in generating and validating scientific knowledge. In the modern world, professional scientists are usually employed by large institutions such as universities or companies, and work as members of teams. Although individual scientists and small groups advance knowledge claims quite regularly, these only attain any significance if other scientists agree that they should be published in journals, or take up the ideas themselves. In this way, scientific communities determine both what is accepted as scientific knowledge and the formalisms and conventions that are used when talking scientifically about the natural world (see Chapters 5 and 6).

Hence, it is not possible for learners to simply 'discover' scientific knowledge for themselves because scientific knowledge is more than a description of how the world works. No amount of 'experimentation' with electric cells and lengths of wire will enable a student to formulate the conventions used to describe magnetic fields and to develop the scientific laws of electromagnetism. Rather than 'finding out for oneself', learning science involves being introduced to, and coming to accept and understand, some of the norms, the ways of thinking, and the ways of explaining used in the scientific community.

It is also the case that the pupils in our schools live within a community that has its own, 'everyday' or 'commonsense' ways of talking and thinking about the events and phenomena which are of interest to scientists. Consider, for example, the case of air pressure. The action of drinking orange through a straw is described in everyday situations in terms of 'sucking'; for example, children from an early age are able to respond to parents' requests to 'suck quietly' as they drink through a straw. For the parent and child, little difficulty is involved in establishing a shared understanding of what is meant, as the child is immersed in this kind of talk from birth.

However, potential difficulties in establishing shared understandings *do* arise when the child revisits these familiar events in school science lessons. It is here, in the classroom, that the teacher faces the challenge of introducing students to the scientific ways of interpreting and explaining phenomena, which the students already think about in their own everyday ways. A three-way relationship thus exists between the phenomenon, the everyday ways of talking about that phenomenon, and the scientific description.

A great deal of research has been carried out internationally into the ways that students typically explain natural phenomena (see, for example, Pfundt and Duit 1990) and how these so-called *alternative conceptions* develop over time (Driver *et al.* 1994b). Given the all-prevailing nature of everyday knowledge, there is a strong argument to suggest that it is the scientific view which

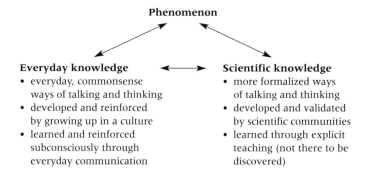

Figure 3.1 Everyday and scientific knowledge

offers an 'alternative' perspective to everyday views (rather than the other way round).

In summary, learning science involves coming to understand, and being able to use, the conceptual tools of the scientific community. In this process the learner is engaged in making sense of the scientific view, and this learning is carried out against a backdrop of existing everyday ways of thinking about the phenomena under scrutiny. This kind of perspective on science learning (and indeed learning in general), which draws attention to the active role of the learner and the interplay between existing and 'new' knowledge, is often referred to as a *constructivist* view of learning (see, for example, Driver and Oldham 1986).

Personal and social perspectives on science learning

Much of the earlier work carried out by 'constructivist researchers' focused on the individual learner, identifying patterns in their existing alternative conceptions about particular phenomena, and also, investigating the conditions required for science learning to occur. Posner *et al.* (1982), in developing their 'Conceptual Change Model' of learning, identified the need for new knowledge to be 'intelligible', 'plausible' and potentially 'fruitful' for the learner if learning, or conceptual change, is to occur. Their focus was very much on changes in individual learners' conceptual frameworks, and hence this kind of perspective on learning is often called 'personal constructivism'.

However, personal constructivist perspectives on learning say little about the *social* features of learning environments, such as interactions between groups of students, or teacher and students, and how these influence learning. More recently there has been a development in science education research towards acknowledging that learning a body of formal knowledge – such as science – inevitably takes place in a social context (probably in a school), and that the social context is highly influential on learning. Joan Solomon (1994) offers a useful metaphor for science learning in which she imagines a child sitting outside the family circle: listening to the words and phrases used, building up ideas, trying out the sense of those ideas with elders, receiving new and helpful explanations, gradually having ideas accepted by others, and being encouraged to use those ideas in new ways. This is a picture of learning which brings to the fore the importance of *talk* in exploring new ideas and learning how to use them in appropriate contexts and in appropriate ways.

A view of learning: including both personal and social perspectives

One theoretical view of learning which has been influential in the development of social perspectives in science education is that of the Russian psychologist Vygotsky (1896–1934). In his 'General genetic law of cultural development', Vygotsky (1978) states that all higher psychological processes and structures (such as science concepts) originate on the social plane. In effect, such processes and structures are first encountered by learners as they listen to the talk of others, or read the writing of others.

From a Vygotskian perspective, social context and language are fundamental to learning. At the same time, however, it should be recognized that internalization cannot simply involve direct transfer of 'ways of talking' from social to internal planes: a step of personal interpretation or personal sense-making is also necessary (Leontiev 1981). In simple terms, individual learners must make sense of the talk which surrounds them, and in doing so, relate it to their existing ideas and ways of thinking: learners must reorganize and reconstruct the talk and activities of the social plane. In this respect Vygotskian theory shares common ground with personal constructivist perspectives in recognizing that the learner cannot be a passive recipient of knowledge.

In effect, the Vygotskian view of learning offers a bringing together of personal and social perspectives on learning. The social origins of learning and the fundamental role of language in learning are recognized, alongside the essential personal sense-making or interpretive step.

The personal sense-making step: the concept of 'learning demand'

As all science teachers are very well aware, learning science often creates difficulties: the learning outcomes which follow from instruction are often disappointing in terms of how much students understand; how much they are able to apply; and how much they are able to remember. Why should learners experience such difficulty?

As outlined earlier, learning science involves developing ways of thinking about phenomena which we already talk about in familiar, everyday ways. Not only is it the case that science offers a *different* way of thinking about phenomena, but the fact is that the scientific view can often appear counter-intuitive, challenging commonsense notions about those phenomena. Lewis

Wolpert (1992) acknowledges this point in referring to the 'uncommonsense of science'. The nature of any differences between everyday and scientific views can be thought of in terms of the *learning demand* (Leach and Scott 1995). The learning demand is a description of the differences between everyday and scientific ways of thinking about the world, and the resultant challenges that learners will face in coming to internalize and understand scientific accounts of phenomena.

The learning demand can relate to various kinds of differences: differences in the conceptual tools used; differences which relate to basic assumptions about the nature of the world (ontological assumptions); and differences which relate to the nature of the knowledge being used (epistemological assumptions). This can be illustrated with our previous example of the air pressure explanation for drinking through a straw. The learning demand for an individual learner might involve: using the concept of 'air pressure' rather than that of 'sucking' (a conceptual demand); coming to accept that air is a substantial material which *can* exert large pressures (an ontological demand); and appreciating that the concept of air pressure is generalizable and can therefore be used to explain a whole range of different phenomena (an epistemological demand). Returning to the ideas of Posner *et al.* (1982), it might well be the case that the learner can understand the air pressure explanation (it is *intelligible*), but they just can't believe it or it doesn't seem to make sense (it isn't *plausible*). In this case, learning the scientific view involves the learner in making significant changes to their fundamental assumptions about the nature of the world ('surely it can't be the air which is pushing the juice up through the straw?').

The concept of learning demand helps focus the teacher's attention on the personal steps required for sense-making in the learning of science, and further, provides a starting point for identifying the nature of any difficulties which the learner is likely to experience in coming to accept the scientific point of view.

In summary, we can say:

- Learning science involves being introduced to the ways of talking and thinking used in the scientific community, which are based on particular scientific concepts and modes of explaining;
- Students already have everyday ways of thinking (alternative conceptions) about the phenomena, and detailed accounts of these alternative conceptions now exist in a number of topic areas;
- Alternative conceptions are continually reinforced in day-to-day talk and are often different to the scientific view; as such alternative conceptions can act as a 'barrier' to science learning;
- Science learning originates in social situations, and language is central to teaching and learning science. Language provides the means by which new ideas are first introduced and rehearsed and also the 'tools' for pupil thinking. The 'talk of science' provides the conceptual tools for 'thinking about science'.

What are the messages from these views of learning for approaches to science education?

There is a big leap between knowing the difficulties that learners face in coming to understand science, and seeing how this might be drawn upon in practice. We would claim that the research evidence on teaching and learning science has useful messages about the practice of science education in the following areas:

- the aspects of scientific subject matter that need to be emphasized in teaching;
- the effectiveness of some teaching approaches in a specific topic area compared with others;
- the teacher's role in explaining scientific concepts so that young people understand them;
- how scientific ideas might be introduced and sequenced in the curriculum to maximize learning.

Following David Ausubel (1968), these points have traditionally been referred to as 'finding out where learners are at' and 'teaching accordingly'.

The aspects of scientific subject matter that need to be emphasized in teaching

The notion of 'learning demand' was introduced earlier as a useful way of identifying what it is about a given scientific idea that needs emphasizing to students in order that scientific explanations are understood. This is perhaps best illustrated by working through an example in rather more detail.

Ideas about plant nutrition are typically introduced to students during secondary schooling. The key ideas presented to students are that plants synthesize carbohydrates from carbon dioxide and water in the process of photosynthesis; that these carbohydrates act both as a substrate for respiration in the plant's cells and as the main raw material in the synthesis of biomass; and that oxygen and water are by-products of photosynthesis. A number of studies have been conducted around the world into how students explain plant nutrition, both prior to any formal teaching and after teaching (see Wood-Robinson 1991 for a review). These studies identify a number of characteristic features of students' reasoning about the topic:

- organisms such as trees and aquatic plants are often not considered to be plants at all by younger students, and their nutritional needs may also be thought of as different;
- 'food' for plants is often thought of as being similar to food for animals in that it is ingested rather than synthesized from simpler molecules;
- nutrients such as minerals and vitamins are often not distinguished by students from the respirable substances produced in photosynthesis;

- the process by which plants increase their biomass is often not considered by students;
- students find it implausible that plant biomass can be synthesized from a gas (i.e. carbon dioxide) and a liquid (i.e. water);
- the processes of photosynthesis, respiration and gas exchange are often not differentiated by students.

Perhaps the most challenging learning demand faced by students in this area concerns the issue of accepting that plant biomass, including something as 'solid' as wood, can be made from a gas in the atmosphere and water. This is an ontological issue: it relates to students' understanding about the nature of the world. There is a good deal of evidence to suggest that many students are happy to accept that matter can 'appear and disappear', and do not therefore see any need to account for the origins and fate of matter in change processes, and especially change processes in living things (for example Andersson 1991; Leach *et al.* 1991, 1996). For students who reason in this way, the goal of teaching must be to build a model in which students view matter in the solid, liquid and gaseous states as fundamentally similar, so that they can appreciate how plant biomass is chemically synthesized from atmospheric carbon dioxide and water. In addition to all of these challenges, teaching needs to help students to differentiate between the biochemical functions of photosynthesis, and respiration and the process of gas exchange.

Traditional approaches to planning teaching involve breaking the subject matter down into clear, logical steps and then thinking how these might best be explained to students. However, some of the learning demands faced by students might never be identified through this approach. For example, in the case of photosynthesis the fundamental necessity for students to come to a view that solid biomass can be synthesized from an atmospheric gas and water might never be recognized, as this model of matter is taken for granted in scientific accounts of photosynthesis.

Once learning demands for topics such as plant nutrition are identified, by drawing upon studies of the everyday knowledge that students tend to use, it becomes possible to design and evaluate new teaching strategies. In the case of photosynthesis, a number of teaching strategies have been proposed which are informed by research (for example CLISP 1987; Barker 1989). Activities within these teaching schemes are selected to address the learning demands identified in a systematic way. For example, in order to address the issue of 'food' for plants being seen of as similar to 'food' for animals, Barker's scheme starts teaching about plant nutrition with the question 'Where does the wood come from?' rather than the normal approach of asking where plants get their food. In this way, differences between the nature of plant and animal 'food' can be discussed explicitly *after* direct teaching about plant nutrition.

But there is more to designing teaching than addressing each learning demand in turn: students need to be taught how key concepts are drawn upon and link together. In teaching about plant nutrition, for example, it is necessary to introduce fundamental ideas about matter and energy, and to show how these are used in explaining plant nutrition, as well as showing how plant nutrition relates to other biological concepts such as transfer of matter and flows of energy in ecosystems. However, there is ample research evidence that *learning* and *teaching* do not proceed at the same pace. For example, few students who follow carefully designed teaching schemes, such as those mentioned above, come to a scientifically accurate view of plant nutrition by the end of teaching (for example Oldham *et al.* 1990). Rather, most students tend to make progress in smaller steps, using notions at the end of teaching that represent an advancement in thinking, but which fall short of the scientific view. For example, in the case of plant nutrition, many students who think that plants get biomass from the soil before teaching, say that plants get biomass from the soil *and photosynthesis* after teaching.

This kind of approach to planning and implementing teaching, based upon identifying learning demands, has now been worked through in many other parts of the science curriculum (for example mechanics – Viennot 1979; heat and temperature – Tiberghien 1985; reaction forces – Brown and Clement 1991; factors influencing rusting – Scott *et al.* 1994; particulate theory of matter – Andersson and Bach 1996; entropy and change – Boohan 1996; the solubility concept – Mirzalar-Kabapinar 1999). Further summaries of research findings about students' everyday knowledge across a broad range of topics typically taught to secondary school science students, with suggestions for teaching approaches, are given by Driver *et al.* (1994a).

The effectiveness of some teaching approaches in a specific topic area compared with others

Few studies have been designed to compare the effectiveness of such teaching with that of more traditional approaches. This is perhaps unsurprising, when one considers what is involved in conducting this sort of experimental study. At the outset, researchers must get access to enough 'control' and 'experimental' classes to make a valid comparison. The students in 'control' and 'experimental' classes have to be comparable in ability, and researchers have to devise ways of measuring this comparability. The quality of the teaching has to be uniform, even though the teaching approach used in control and experimental classes will be different. This problem is far from trivial. For example, it would not be useful to compare students' performance in control classes taught by Mr X, who is widely perceived as boring by his students, with the performance of students in the experimental class taught by the dashing and dynamic Ms Y, as any differences could be due to the students' reactions to the teacher rather than to the teaching approach. A solution might be to get Ms Y to teach both control and experimental classes, but if Ms Y believes in the experimental teaching approach, and has developed her own expertise through being involved in a research project, it may not be possible for her to teach the control class in a 'normal' way. And finally, researchers have to decide the most appropriate way of assessing students' learning. For example, although control and experimental classes might perform similarly on typical examination questions, the long-term retention of control and experimental classes might well be different.

In spite of these methodological difficulties, a small number of experimental studies have been conducted to evaluate the effectiveness of teaching approaches based on addressing specific learning demands. One of these involves teaching about gravity and inertia (Brown and Clement 1991). For many learners, 'force' is associated with movement. It is therefore very difficult for such learners to accept that an object at rest, such as a mug of coffee on a table, is subject to a gravitational force and to an equal and opposite force from the table on the mug. Typically, students think that no upward forces are acting when the mug is at rest. A teaching sequence was prepared to address this learning demand, using what Brown and Clement term 'bridging analogies'. Bridging analogies are designed to encourage students to draw an analogy between an intuitively obvious idea, and a counter-intuitive one. In this case, Brown and Clement suggested placing a heavy object upon a student's upturned hand. Students readily accepted that their hand exerted an upward force on the object as they felt the weight of the object on their hand and they needed to push up to balance the downward force. Then, the same object was placed on a long ruler, which was supported at each end but not in the middle. As they can see the ruler bend under the object's weight, students were persuaded that the ruler, by virtue of its springiness, exerts an upward force on the object. Finally, students were asked to make a logical leap from the hand and ruler examples to the original target situation of an object on a table. The notion of there being an upward force from the table (due to the springiness of the table) onto the mug was then introduced.

The experimental teaching sequence was used with 79 students. The effectiveness of this teaching was compared with 132 students in control groups. All students completed the same test items prior to, and after, teaching. The post-test measured gains after a period of 2–4 months. The experimental group showed improvements in post-test scores in the order of 27–29 percentage points, compared to the control group.

A common criticism of teaching approaches based upon an analysis of learning demands is that they are unfeasible because they are time consuming. However, in a recent study of upper secondary students learning about electricity in physics, Laurence Viennot and Sylvie Rainson (1999) present compelling evidence that such teaching sequences *which take no more time than conventional approaches* can be designed, and furthermore, that these sequences result in consistent improvements in student learning compared with traditional approaches. In this study, the same teacher taught 'control' classes, using a conventional teaching sequence, and 'experimental' classes using the designed teaching sequence, thereby addressing, at least to some extent, the problem of comparing an enthusiastic teacher's use of experimental teaching approaches with a less enthusiastic teacher's use of conventional methods. In addition, the students in the 'control' classes in the study would have been expected to perform better than those in 'experimental' classes, given their previous higher academic performance. The teaching approach, used over four years, showed that mean student test scores in the 'experimental' classes were typically in the order of 20 per cent better than those for students in the 'control' classes and were very stable. This suggests that improvements in student learning were due to the teaching approach used rather than teacher effects.

Other smaller-scale studies have been conducted (for example Johnston and Driver 1990; Mirzalar-Kabapinar 1999), which tend to show significant but fairly small improvements in students' learning as a result of following teaching interventions designed to address specific learning demands. In summary, there is modest but persuasive evidence to suggest that teaching approaches, which are based upon an analysis of learning demands, are more effective at promoting learning than are conventional approaches. However, further large-scale investigations of the effectiveness of teaching approaches based on the analysis of learning demands, of the kind carried out by Philip Adey and Michael Shayer (1993) in the context of teaching for cognitive acceleration, would be valuable in determining the effectiveness of such approaches.

The teacher's role in explaining scientific concepts so that young people understand them

Recent development of interest in social perspectives on science learning has led to an alternative focus for science education research: one which acknowledges the importance of teaching activities but focuses rather on the teacher and pupil *talk* which surrounds those activities. The particular orientation of this research into teacher and student talk is towards how scientific ideas are made available by the teacher in the classroom and how shared meanings are developed between teacher and students.

A frequently cited and important study in this area of research is the work of Edwards and Mercer (1987), presented in the book *Common Knowledge*. From their analyses of classroom discourse, Edwards and Mercer identify how the teacher controls the teaching and learning events of the classroom, maintaining a tight definition of what became joint versions of events, and joint understandings of curriculum content (Edwards and Mercer 1987). From their findings they develop a typology of the kinds of interventions made by teachers to achieve this control. Typical teacher interventions include, for example:

- marking knowledge as significant and joint – where expressed knowledge is given special prominence by discursive practices such as special enunciation, repetition and the use of formulaic phrases;
- cued elicitation of pupils' contributions – where the teacher asks questions while simultaneously providing explicit clues to the information required, a process which may be accomplished with the help of teacher intonation, pausing, gestures or physical demonstrations;

- paraphrasing pupils' contributions and offering reconstructive recaps – where the teacher reinterprets what pupils have said in order to maintain a strict control over the content of developing common knowledge.

Through such interventions, the teacher is able to work with whole classes of pupils to make the 'scientific story' available and prominent against a background of everyday ways of talking about and explaining the phenomenon under consideration. Edwards and Mercer do not address science learning specifically in their book.

In *Talking Science*, Lemke (1990) points to various shortcomings in the 'traditional' approaches to teaching and learning science which he had observed and offers alternative pedagogical strategies. Lemke argues that scientific reasoning is learned, 'by talking to other members of the community, we practice it by talking to others, and we use it in talking to them, in talking to ourselves, and in writing and other forms of more complex activity' (Lemke 1990: 122).

Consequently, some recent research initiatives have attempted to develop science teaching to maximize the amount of time spent 'talking science' by students, and to take seriously the notion of immersing science pupils in 'discourse communities' (as suggested by Lemke). An approach that has been encouraged in schools in the USA, through current North American curriculum reform recommendations (AAAS 1989; National Research Council 1996), involves students in 'doing' science themselves. In this case, 'doing science' is taken as students identifying problems, framing questions and working with the teacher as a consultant to talk through, and to develop, possible solutions. This field of curriculum development and research draws on the principles of situated cognition (see Brown *et al.* 1989; Rogoff 1990; Lave and Wenger 1991) and has had considerable impact in both science education (see, for example: Eichinger *et al.* 1991; Moje 1995; Roth 1996; Roychoudhury and Roth 1996) and mathematics education (see, for example: Lampert 1990; Cobb *et al.* 1998).

In all of the studies cited here students are directly involved in 'doing science' or 'doing mathematics'. Roychoudhury and Roth (1996), for example, investigate interactions in an open-inquiry physics laboratory involving junior high school students. They describe the open-inquiry laboratory as being one in which, 'the activities are open-ended . . . there is no recipe-type, step-by-step procedure available for conducting the experiments'; the students, 'have decision-making power over what to investigate and how to investigate within the constraints of available resources'.

Ogborn *et al.* (1996) combine their interests in the language of the science classroom with a focus on traditional teacher-led classrooms by examining how high school science teachers construct and present *explanations* in the classroom. The authors offer as the main outcome of the research a way of thinking about explanations, and present a theoretical framework which has three main components (Ogborn *et al.* 1996):

1 scientific explanations as analogous to stories;
2 an account of meaning-making in explanation consisting of four main
  parts:
    – creating differences
    – constructing entities
    – transforming knowledge
    – putting meaning into matter;
3 variation and styles of teacher explanation.

In developing the notion of scientific explanations as being analogous to
stories, Ogborn *et al.* suggest that the vital features of a scientific story are
that: first there is a cast of protagonists, each of which has its own capabilities
which are what makes it what it is (protagonists might include entities such
as electric currents, germs, magnetic fields, and also mathematical construc-
tions such as harmonic motion and negative feedback); secondly that the
members of this cast enact one of the many series of events of which they
are capable; and that lastly these events have a consequence which follows
from the nature of the protagonists and the events they happen to enact.

Ogborn *et al.* maintain that the first step in meaning-making in explanation is to create a need for that explanation and suggest that this can be
achieved through identifying a difference to be bridged or resolved. In the
classroom, creating differences might be achieved through: promises of clarification; eliciting differences of opinion; using stories to suggest ideas; dis-
playing counter-intuitive results and creating expectations.

Having created an opening for the explanation, the teacher is then faced
with the fact that the worlds of protagonists, which constitute scientific
explanations, are often far from everyday commonsense (a point made earlier in this chapter), and that scientific explanations can make no sense to
learners until they know what the entities involved are supposed to be able
to do or have done to them. Thus there is the need for students to construct
explanatory entities or there is the need to 'talk into existence' these entities.
The process of talking into existence explanatory entities involves personal
transformation of meaning- and sense-making by students. In addition to
this personal transformation of knowledge, Ogborn *et al.* point out the ways
in which knowledge is transformed as it is presented in the classroom –
pointing to the use of narrative and the crucial role of analogy and metaphor
(with, for example, the eye seen as a camera and the control of the hormone
system by the pituitary gland seen as a conductor keeping an orchestra
together).

All of the studies referred to in this section share a common theme in
focusing attention on the patterns in teacher talk, and the crucial importance
of that talk in supporting student meaning-making (see Chapter 6). So, in
summary, what can we point to as the messages coming out of this area of
research into language, teaching and learning in science? Two important
general points are, first, the ways in which this research has helped to
re-establish interest in the key role of the science teacher (after 20 years

of research into pupils' thinking), and secondly, the ways in which it has
re-focused attention on the crucial importance of teacher and pupil talk
in learning science (following a period of pre-occupation with teaching
activities).

How scientific ideas might be introduced and sequenced in the curriculum to maximize learning

Traditional approaches to writing science curricula involve getting groups of
science education experts together to make recommendations based upon
their own professional experience. However, we would argue that, in addition, the notion of learning demand can be useful in informing decisions
about the age at which ideas might be introduced to young people through
the science curriculum, and the sequence in which ideas should be introduced (for a fuller treatment, see Driver *et al.* 1994c). Here, the implications of
research on teaching and learning science are more directly relevant to those
responsible for curriculum monitoring and development than to science
teachers themselves.

This issue can be illustrated with reference to the National Curriculum in
England and Wales. A key feature of the National Curriculum introduced in
1989 (DES 1989) was that the expected development of students' understanding of specific ideas was mapped out between the ages of 5 and 16 over
a ten-level scale. For example, in the area of plant nutrition the youngest
pupils (aged 5 to 7) would learn about the basic needs of plants by taking
care of them, whereas in the 14–16 age range young people would learn
about the process of photosynthesis and its role in matter cycling and energy
flow in ecosystems. Since 1989, the National Curriculum has been revised
twice. On each occasion, those responsible for the revisions commissioned
science education researchers to give advice about decisions concerning the
age placement and sequencing of concepts in the curriculum (for example
Leach *et al.* 1991; Adey *et al.* 1994a, 1994b). In order to give advice, research
on the everyday knowledge of young people of a given age was compared
with curriculum requirements, and learning demands were identified. In
cases where the introduction of ideas to students at a given age involved significant learning demands, it is likely that teaching would not result in effective learning. For example, Leach *et al.* (1991) present evidence that children
of primary age rarely use models of matter capable of explaining the role of
processes such as decay in ecosystems, whereas this idea was included in the
1989 National Curriculum for primary age pupils. Here, it was recommended
that the role of decay in matter cycling should not be introduced until a later
stage.

In other cases, it appeared that important learning demands were not
addressed specifically through the curriculum. For example, Leach *et al.*
(1991) present evidence that many students in the early years of secondary
schooling do not understand competition processes within, and between,
species. Indeed, many students do not appear to think about organisms

as members of populations or ecosystems, focusing instead upon individuals in isolation. The nature of competitive relationships in ecosystems was not addressed in any detail through the curriculum, and the introduction of additional content was therefore recommended. In such ways, research is able to make a further contribution to the practice of science teaching.

Conclusions

The process of teaching and learning science is complex. It certainly cannot be reduced to a matter of selecting a few key variables and following a neat and well-tested algorithm. This is why the variety of approaches used by different teachers often appear to have similar outcomes. It is therefore important for those with an interest in research on teaching and learning science – whether teachers, funders, politicians or professional researchers – to be realistic in their aspirations for research. If the criterion for success of educational research is providing a quick 'technical fix' to a given problem, research on teaching and learning science will always be judged to have failed. By contrast, if the aim of research on teaching and learning is viewed more broadly as clarifying the learning goals which teachers and curriculum developers have for students, developing teaching strategies to address those goals, and obtaining feedback from students to determine whether the pedagogical strategies adopted have been successful (Driver 1997), then we think that this chapter shows that it becomes possible to identify existing successes and future research priorities.

References

AAAS (1989) *Science for all Americans: Project 2061*. Washington, DC: AAAS.
Adey, P. and Shayer, M. (1993) *Really Raising Standards*. London: Routledge.
Adey, P., Asoko, H., Black, P. *et al.* (1994a) 1994 revision of the National Curriculum for science: implications of research on children's learning in science. Unpublished report to SCAA. King's College London.
Adey, P., Asoko, H., Barker, J. *et al.* (1994b) Science in the National Curriculum: advice to SCAA on the draft proposals (May 1994). Unpublished report, King's College London.
Andersson, B. (1991) Pupils' conceptions of matter and its transformations (age 12–16), *Studies in Science Education*, 18: 53–85.
Andersson, B. and Bach, F. (1996) Developing new teaching sequences in science: the example of 'bases and their properties, in G. Welford, J. Osborne and P. Scott (eds) *Research in Science Education in Europe: Current Issues and Themes*. London: Falmer.
Ausubel, D.P. (1968) *Educational Psychology: A Cognitive View*. New York: Holt, Rinehart and Winston.
Barker, M. (1989) Teaching and learning about photosynthesis 2: a generative learning strategy, *International Journal of Science Education*, 11(2): 141–52.

Boohan, R. (1996) Using a picture language to teach about processes of change, in G. Welford, J. Osborne and P. Scott (eds) *Research in Science Education in Europe: Current Issues and Themes*. London: Falmer.

Brown, D. and Clement, J. (1991) Classroom teaching experiments in mechanics, in R. Duit, F. Goldberg and H. Niedderer (eds) *Research in Physics Learning: Theoretical and Empirical Studies*. Kiel, Germany: IPN.

Brown, J.S., Collins, A. and Duguid, P. (1989) Situated cognition and the culture of learning, *Educational Researcher*, 21(5): 31–5.

CLISP (1987) *Approaches to Teaching Plant Nutrition*. Leeds: Centre for Studies in Science and Mathematics Education.

Cobb, P., Perlwitz, M. and Underwood-Gregg, D (1998) Individual construction, mathematical acculturation, and the classroom community, in M. Larochelle, N. Bednarz and J. Garrison (eds) *Constructivism and Education*. Cambridge: Cambridge University Press.

DES (1989) *Science in the National Curriculum*. London: HMSO.

Driver, R. (1997) The application of science education theories: a reply to Stephen P. Norris and Tone Kvernbekk, *Journal of Research in Science Teaching*, 34(10): 1007–18.

Driver, R. and Oldham, V. (1986) A constructivist approach to curriculum development in science, *Studies in Science Education*, 13: 105–22.

*Driver, R., Asoko, H., Leach, J., Mortimer, E. and Scott, P. (1994b) Constructing scientific knowledge in the classroom, *Educational Researcher*, 23(7): 5–12.

Driver, R., Leach, J., Scott, P. and Wood-Robinson, C. (1994c) Young peoples' understanding of science concepts: implications of cross-age studies for curriculum planning, *Studies in Science Education*, 24: 75–100.

*Driver, R, Squires, A., Rushworth, P. and Wood-Robinson, V. (1994a) *Making Sense of Secondary Science*. London: Routledge.

*Edwards, D. and Mercer, N.M. (1987) *Common Knowledge: The Development of Understanding in the Classroom*. London: Methuen.

Eggleston, J., Galton, M. and Jones, M. (1976) *Processes and Products of Science Teaching*. London: Macmillan.

Eichinger, D.C., Anderson, C.W., Palincsar, A.S. and David, Y.M. (1991) An illustration of the roles of content knowledge, scientific argument, and social norms in collaborative problem solving. Conference paper, Chicago annual meeting of AERA.

Johnston, K. and Driver, R. (1990) *A Case Study of Teaching and Learning about Plant Nutrition*. Leeds: Centre for Studies in Science and Mathematics Education.

Lampert, M. (1990) When the problem is not the question and the solution is not the answer: mathematical knowing and teaching, *American Educational Research Journal*, 27: 29–64.

Lave, J. and Wenger, E. (1991) *Situated Learning: Legitimate Peripheral Participation*. New York: Cambridge University Press.

*Leach, J. and Scott, P. (1995) The demands of learning science concepts: issues of theory and practice, *School Science Review*, 76(277): 47–51.

Leach, J.T., Driver, R.H., Scott, P.H. and Wood-Robinson, C. (1991) Progression from age 5 to age 16 about cycles of matter, flows of energy and interdependency and classification of organisms in ecosystems. Unpublished report to the National Curriculum Council, University of Leeds.

Leach, J.T., Driver, R.H., Scott, P.H. and Wood-Robinson, C. (1996) Children's ideas about ecology 2: ideas about the cycling of matter found in children aged 5–16, *International Journal of Science Education*, 18(1):19–34.

*Lemke, J.L. (1990) *Talking Science: Language, Learning and Values*. Norwood, New Jersey: Ablex Publishing Corporation.

Leontiev, A.N. (1981) The problem of activity in psychology, in J.V. Wertsch (ed.) *The Concept of Activity in Soviet Psychology*. Armonk, NY: Sharpe.

Mirzalar-Kabapinar, F. (1999) Teaching for conceptual understanding: developing and evaluating Turkish students' understanding of the solubility concept through a specific teaching intervention. Unpublished PhD, School of Education, The University of Leeds.

Moje, E.B. (1995) Talking about science: an interpretation of the effects of teacher talk in a high school classroom, *Journal of Research in Science Teaching*, 32(4): 349–71.

National Research Council (1996) *National Science Education Standards*. Washington, DC: National Academy Press.

*Ogborn, J., Kress, G., Martins, I. and McGillicuddy, K. (1996) *Explaining Science in the Classroom*. Buckingham: Open University Press.

Oldham, V., Driver, R. and Holding, B. (1990) *CLIS Teaching Schemes in Action: Case Studies on Plant Nutrition*. The University of Leeds: Centre for Studies in Science and Mathematics Education.

Pfundt, H. and Duit, R. (1990) *Bibliography: Students' Alternative Frameworks and Science Education*. Kiel, Germany: IPN.

Posner, G.J., Strike, K.A., Hewson, P.W. and Gerzog, W.A. (1982) Accommodation of a scientific conception: toward a theory of conceptual change, *Science Education*, 66(2): 211–27.

Rogoff, B. (1990) *Apprenticeship in Thinking: Cognitive Development in Social Context*. Oxford: Oxford University Press.

Roth, W.-M. (1996) Teacher questioning in an open-inquiry learning environment: interactions of context, content and student responses, *Journal of Research in Science Teaching*, 33(7): 709–36.

Roychoudhury, A. and Roth, W.-M. (1996) Interactions in an open-inquiry physics laboratory, *International Journal of Science Education*, 18(4): 423–45.

Scott, P., Asoko, H., Driver, R. and Emberton, J. (1994) Working from children's ideas: planning and teaching a chemistry topic from a constructivist perspective, in R. Gunstone and R. White (eds) *The Content of Science*. London: The Falmer Press.

Solomon, J. (1994) The rise and fall of constructivism, *Studies in Science Education*, 23: 1–19.

Tiberghien, A. (1985) Heat and temperature, the development of ideas with teaching, in R. Driver, E. Guestne and A. Tiberghien (eds) *Children's Ideas in Science*. Buckingham: Open University Press.

Viennot, L. (1979) Spontaneous reasoning in elementary dynamics, *European Journal of Science Education*, 1(2): 205–21.

Viennot, L. and Rainson, S. (1999) Design and evaluation of a research-based teaching sequence: the superposition of electric field', *International Journal of Science Education*, 21(1): 1–16.

Vygotsky, L.S. (1978) *Mind in Society: The Development of Higher Psychological Processes*. Cambridge, MA: Harvard University Press.

Wolpert, L. (1992) *The Unnatural Nature of Science*. London: Faber and Faber.

Wood-Robinson, C. (1991) Young people's ideas about plants, *Studies in Science Education*, 19: 119–35.

4 The role of practical work

Rod Watson

Empirical work is one of the defining features of science. Many countries devote considerable resources to give students of science the opportunity of doing practical work in their science lessons (Beatty and Woolnough 1982; Watson and Prieto 1994). But does it work? Is it worth the investment? Can it be used more effectively?

Many scientists and science educators are convinced that practical work *must* play an important role in learning science, but the reasons for its prominence are less clear. This lack of clarity lies in the vagueness of the questions asked about the role of practical work. Asking about the effectiveness of practical work for learning is like asking whether children learn by reading. The answer lies in the nature and contents of the activities and the aims which they are trying to achieve. Just as there is a great variety of styles and contents of reading matter, there is also a great variety of kinds of practical work. A more focused question is to ask what kinds of practical activities can be used to achieve particular aims. Practical work may be used in a variety of formats such as practicals following recipes, investigations, skills training, teacher demonstrations to promote discussion about phenomena and raise questions, problem-solving activities, and heuristic practical activities designed to help students induce generalizations.

This chapter looks at practical work – its aims, its effectiveness, its use for teaching conceptual aims, and the strategies for the development of practical procedural knowledge and its use in investigations.

Aims: What is practical work? What is it for?

Several studies have collected teachers' views on the aims of practical work. Of particular interest are the studies of changes over time. Ten aims in the Kerr (1964) survey were augmented by a further ten in the Beatty and Woolnough (1982) survey. That set of 20 aims was again used in the Swain *et al.* (1998) study. Despite changes in the kinds of practical work done over time, in all three studies four aims remained the most popular:

- to encourage accurate observation and description;
- to make phenomena more real;
- to arouse and maintain interest;
- to promote a logical and reasoning method of thought.

There is, however, a cluster of aims that were rated more highly in the Swain *et al.* study in 1998 than those in the Beatty and Woolnough study in 1982. They are:

- to practise seeing problems and seeking ways to solve them;
- to develop a critical attitude;
- to develop an ability to cooperate;
- for finding facts and arriving at new principles.

Following the introduction of more open investigational work in the National Curriculum in England, this change in emphasis of aims is a product of a change in the kind of practical work used.

The messages here are clear. Teachers see both procedural and content aims as part of the core of practical science and as inextricably related to one another. For example, in order 'to encourage accurate observation and description' one has to observe some phenomenon and, reciprocally, accurate observation and description aids in making phenomena more real. Beyond these core aims are different sets of aims associated with different kinds of practical work. The difficulties of trying to achieve too many aims through one activity are discussed by Woolnough and Allsop (1985). They argued for different kinds of practical activity for different aims: that is, short illustrative tasks to stimulate discussion and learning about concepts; practical activities to develop practical skills and procedures; and more extended and open practical experiences to develop investigative skills.

The effectiveness of practical work in general

Hodson (1993) has reviewed the effectiveness of practical work under four headings: motivation, acquisition of skills, learning scientific knowledge and the methods of science, and scientific attitudes. The results of his study indicate that, in each of these areas, school practical work leaves much to be desired.

On the whole, pupils enjoy practical work and develop positive attitudes to it, but this enjoyment is qualified. A significant minority of pupils express a dislike for practical work (Head 1982), and enthusiasm for practical work often declines with age (Lynch and Ndyetabura 1984). What appears to be important is not whether pupils do practical work but the kinds of practical work used. Open kinds of practical work are seen by teachers as very motivating – motivation is improved if pupils feel a sense of ownership of investigations (Kempa and Dias 1990; Jones *et al.* 1992) and greater control is given to pupils.

Several studies have shown variable success in performing practical tasks (for example APU 1982, 1985; Gott and Duggan 1995: Chapter 5). The TIMSS (Third International Mathematics and Science Study) shows great variability between countries, with grade 8 students in Singapore and England performing significantly better (Harmon *et al.* 1997) than students in other countries, on practical tasks designed to test skills such as:

- measuring; the use of simple experimental and mathematical procedures;
- designing and implementing approaches to solve problems or investigate phenomena;
- synthesizing knowledge, application, and personal experience into an interpretation of the data.

There is also variability between different skills and process. For example, in the APU study (1985) 15-year-olds were asked to read pre-set values on several simple measuring instruments. Fewer than one in five correctly read an ammeter and only about half correctly read the value on a rule. Performance was better in reading a thermometer and a force meter. In 'making and interpreting observations', however, about 50 per cent of pupils performed simple observations successfully. Planning investigations was a strength of pupils in the APU (1982) study, and 15 years later an analysis of the TIMMS data (Harmon *et al.* 1997) shows that this has been sustained.

Hodson (1993) reported that the research literature indicates that there is little to show that practical work is effective in helping students to learn scientific knowledge, and that some reports suggest that it is less successful than other methods. A recent study by Watson *et al.* (1995) gave typical results. The understandings of two groups of 150 15-year-old pupils were compared. One group had been exposed to a curriculum with a high practical content (in England) and the other group with a low practical content (Spain). In spite of having substantially more practical experience with combustion, the English sample showed few differences from the Spanish sample in either their scientific or naive conceptions about combustion. What may be more important than the quantity of practical work is what use is made of it in helping pupils to develop scientific concepts. Hodson (1993) reports a similarly disappointing picture with regard to learning about the nature of the methods of science.

He also challenges the efficacy of practical work in teaching supposedly scientific values such as taking a value-free stance, being objective, open-minded and willing to suspend judgement. In contrast, school practical science is dominated by the need to get *correct answers* and find out what *ought to happen*; to ensure conformity with the answer in the textbook. In such conditions it is difficult to see how such attitudes can be generated (Hodson 1993). Mahoney (1979) maintains that scientists frequently display different characteristics anyway: they are often illogical in the way they work, highly selective in reporting data, and that they will stick tenaciously to their views even in the face of contradictory evidence. This brings into question the value of trying to teach such values when recent scholarship in the philosophy and sociology of science casts doubt on how representative such values are. In consequence, Hodson concludes that there needs to be a fundamental rethink of the role of practical work in school science.

Turning from a review of the effectiveness of practical work, this chapter now examines research that explores how practical work may be improved in two broad areas: developing understanding of scientific concepts and developing the capability to carry out scientific investigations. This next section reviews how practical work may be used to develop conceptual understanding and the rest of the chapter then focuses on what is meant by scientific investigations and how pupils can be taught to investigate.

How practical work can be used to achieve conceptual aims

'I listen and I forget, I look and I remember, I do and I understand.' This so-called Chinese proverb summarized the discovery learning approaches of the Nuffield science projects of the 1960s and 1970s. When reflecting on the work on students' understanding of scientific concepts, Ros Driver reformulated the proverb as 'I listen and I forget, I look and I remember, I do and I am even more confused!'

There is a belief among science teachers that practical experience of phenomena is essential for understanding scientific concepts. Recall of incidents, or episodes (White 1988: 31), in the laboratory gives a dimension to scientific concepts that cannot be achieved simply by talking about them. However, progression from observations of phenomena to the construction of scientific concepts is not a simple one. Scientific concepts and theories are often counter-intuitive and have to be constructed in the classroom by talking or reading about phenomena as well as by seeing them. Here, two case studies have been selected from the large number of studies in this area to illustrate the importance of the teacher's role in helping students to construct conceptual understanding in the classroom. One shows the use of a demonstration within a social constructivist approach to learning, and the other the use of practical work in a fairly traditional classroom.

Scott and Leach (1998) describe a teaching episode in which the teacher is building on the idea that reducing the amount of air in a fixed space reduces the air pressure in that space. The teacher sets up a demonstration with two partially inflated balloons inside a bell jar. The air is then sucked out of the bell jar using a vacuum pump and the pupils see the balloons slowly inflating. The pupils are entertained by this unusual sight and start offering their explanations. The scientific explanation of these observations does not flow naturally from simply seeing the demonstration. Pupils are used to the everyday idea that sucking on a straw, sucks up the drink and tend to use a similar explanation that the air is *sucked out* of the bell jar and *sucks* out the walls of the balloon. The teacher's role is to help pupils to understand and use the scientific explanation. To do this, the teacher focuses on key aspects of the demonstration, selecting and emphasizing particular aspects of pupils responses: he points out that the balloons are tightly sealed (that is, the quantity of air in them is fixed); he selects a response from one boy who focuses on the decrease in pressure in the bell jar, praises him for his explanation and then repeats in a slow and deliberate voice, 'So if we make *less* air in the jar there's *less* air pressure in the jar . . .'. What this shows is that the teacher is crucial in introducing a new way of talking about the phenomenon, relating *selected* observations to a scientific explanation. Observations alone are insufficient.

The importance of the teacher as a mediator and interpreter of the observed physical phenomena is also seen in a second example – a case study described by Roth *et al.* (1997) and McRobbie *et al.* (1997). These two papers describe different aspects of six weeks of observation of a physics teacher and his grade 12 students, doing practical work in groups. Different pupils responded to the lessons differently (McRobbie *et al.* 1997). The responses were also different from what the teacher anticipated (Roth *et al.* 1997). The teacher expected students would see the phenomena studied in the same way as he saw them, whereas the students came to the practical activity with their own ideas about motion, and this affected what they observed (Roth *et al.* 1997). The teacher also expected that his instructions were self-evident, but failed to realize that the students did not share his theoretical perspective, which made sense of the practical activities. The result was confusion for the students in understanding what they were supposed to do. The role of the teacher in helping students to construct the accepted scientific view is illustrated in the variable response of the students. Students who felt able to ask many questions were able to use the information provided to come to an understanding of the accepted scientific view (McRobbie *et al.* 1997), whereas the others were left struggling with their own interpretations of the phenomena.

Learning to investigate

What are investigations?

Teachers working in England and Wales have identified the following two characteristics of investigations.

- In investigative work pupils have to make their own decisions either individually or in groups: they are given some autonomy in how the investigation is carried out.
- An investigation must involve pupils in using procedures such as planning, measuring, observing, analysing data and evaluating methods. Not all investigations will allow pupils to use every kind of investigational procedure, and investigations may vary in the amount of autonomy given to pupils at different stages of the investigative process.

(Watson and Wood-Robinson 1998: 84)

The same kinds of justifications have been used for including investigations in the curriculum as those used for practical work in general. The major aim for investigations for many teachers is to develop the use of the procedures of science (Watson and Wood-Robinson 1998) with teaching for conceptual understanding taking second place. A third possible aim is to develop pupils' understanding of the relation between empirical data and scientific theory (Driver *et al.* 1996).

The APU (1982) envisaged investigation as a cyclical process in which students worked through different stages of an investigation: problem generation and perception, reformulation, planning, carrying out practical work by making observations and measurements, recording data, interpreting, evaluating the various preceding stages, and finally, reaching a solution. Investigations were seen as processes which brought together students' conceptual understanding, scientific skills and processes to solve a problem. The work of the APU was concerned with assessing student achievement in specific domains. So, in order to focus on scientific skills and processes, assessment tasks were designed which made low demands on pupils' knowledge and understanding. However, many of these investigations have become incorporated in current curricula, resulting in an artificial separation between experimental/investigative work, and the knowledge and understanding components of the curriculum.

The APU model also placed a heavy emphasis on relations between variables leading to the domination of 'fair testing' in UK curricula. Fair testing describes investigations where an independent variable is manipulated in order to have an effect on a dependent variable, while controlling all other relevant variables. The range of kinds of investigations which fit easily within this model has been criticized as being limited as they fail to represent the variety of methods used by scientists and over-emphasize fair testing

(Watson *et al.* 1998). At ages 11–14 in England and Wales, over 80 per cent of all investigations are fair tests. This means that, if the habitat of animals such as woodlice is being explored, it is much more likely to be done in the laboratory using choice chambers (fair testing), than outside in the natural environment (pattern seeking). Other kinds of investigations such as classifying; identifying; pattern seeking; exploring; investigating models; and making things and developing systems are rarely used (Watson *et al.* 1998, 1999a and 1999b). As well as being dominated by fair testing, the variety of investigations within the fair testing category tends to be very restricted. For example, 16 per cent of all fair testing investigations done at age 11–14 were investigations of variables affecting solubility or rate of dissolution (usually of sugar)!

Teaching and learning scientific procedures

There has been much discussion about the different ways in which scientists go about investigating nature (see Chapter 5). It is now commonly agreed that there is no one scientific method but that scientists work in a variety of ways (Millar and Driver 1987). Scientific skills and processes – the procedures of science – are, however, applied within a variety of different contexts and investigations. But, although the ways in which they are utilized may vary, there are still some procedures common to many kinds of investigations. It is to these procedures that we now turn.

The APU (1988) developed test items for different scientific activities under the headings:

- use of graphical and symbolic representation;
- use of apparatus and measuring instruments;
- observation;
- interpretation and application;
- planning of investigations; and
- performance of investigations.

National surveys of pupils performance were carried out at ages 11, 13 and 15. The findings are reported in four series of reports: reports for teachers (for example Gott and Murphy 1987), research reports (for example APU 1982), reviews of findings (for example APU 1985) and reports looking in more depth at specific skills of process (for example Strang *et al.* 1991). These reports paint a fascinating picture of pupils performance, giving detailed information about progression.

However, one question that remains unanswered in the APU surveys is the extent to which the test items measure skills and processes that can be transferred to other contexts, and to what extent they are measuring skills and processes that are situated in the particular context of the test item:

differences in performance . . . cannot satisfactorily be disentangled from the impact of other aspects of questions, particularly the apparent linguistic demand. Moreover, even within groups of questions . . . collective performance varies markedly, indicating the impact of factors which are unidentified . . .

(APU 1988: 93)

One effect – that of context – was studied by constructing equivalent test items in everyday and school science contexts (Song and Black 1991, 1992). Its effect on performance varied according to the skill being tested, but often the everyday context seemed to cue students into a less scientific way of working.

Whether processes are situation specific or transferable is a question also addressed by Millar and Driver (1987). They argue that the commonly cited 'processes of science' cannot be divorced from the content and context, which actually give meaning to so-called 'process-based' activities in science. This argument is illustrated using the processes of observing, classifying and hypothesizing to show that these only become scientific processes when being used for a scientific purpose. In other words, scientific processes and scientific knowledge and understanding are inextricably linked.

The Procedural and Conceptual Knowledge in Science (PACKS) Project (Millar *et al.* 1994; Lubben and Millar 1996) explored the influence of procedural and conceptual knowledge on pupils' performance in investigative tasks and, in so doing, developed a model of procedural understanding.

A central feature of the PACKS model is that procedural understanding . . . is a knowledge-based domain. It is similar to other science 'content' domains, in that children come to it with prior ideas, and that these ideas may need to be developed or changed through teaching. The domain of scientific evidence contains ideas which must be taught . . .

(Millar *et al.* 1994: 245)

Those ideas which *must be taught* – what they term 'concepts of evidence' – are described as comprising those associated with:

- *design*: variable identification, fair test, sample size, variable types;
- *measurement*: relative scale, range and interval, choice of instrument, repeatability, accuracy;
- *data handling*: tables, graph type, patterns, multi-variate data.

In the PACKS model there is a shift in emphasis compared with the APU, with much more stress being placed, in the former, on understanding the quality of evidence collected rather than on carrying out a particular process. For example, in measuring, the APU focuses on using measuring instruments whereas the PACKS model focuses on 'understanding the appropriate degree of accuracy that is required to provide reliable data which will allow meaningful interpretation.'

This re-definition of skills and processes as procedural understanding leads to the conclusion that there is now 'a need to devise activities which progressively develop and refine children's understanding of the purpose of scientific investigation, and of the key concepts which underpin judgements about the quality of data' (Lubben and Millar 1996). This is similar to the conclusion reached in the 'content' area: it is not sufficient simply to use scientific processes in carrying out a practical activity, but there is also a role for the teacher in helping students to understand the underlying concepts of evidence. Curriculum materials have now been designed to teach such concepts (Foulds *et al.* 1997, 1998).

Teaching and learning in whole investigations

Pupils' interpretation of investigations

The learning of scientific procedures in separate activities and in whole investigations is different. Whole investigations provide a context in which there is a scientific purpose. But, do pupils see this scientific purpose? Research indicates that many do not. Kuhn (1989) proposed a sequence in the development of children's understanding of the relation between theory and data. In the early stages, theory and data are fully integrated and are used interchangeably: children see the purpose of collecting experimental data as being to discover how the physical world works. For example, some young children, when asked to explain what has happened to the water in a puddle that has dried up, will say that it has disappeared. For them, the phenomenon and the explanation are one and the same thing.

Later, when theory and data are compatible, pupils tend to mould them together as 'the way things are', but, when theory and data are incompatible, conflict is avoided by strategies such as selective attention to data. Only in the upper levels of the developmental spectrum are theory and data consciously differentiated. For example, pupils who observe copper turning black on heating are able to consider different theoretical explanations for the phenomenon: the heat may have changed the copper to carbon, or the heating may have caused copper to combine with oxygen to form a black product. Here the theory is clearly separated from the data. Driver *et al.* (1996) used a similar framework distinguishing reasoning about phenomena, reasoning about relationships and reasoning about models in children of three ages. The results indicate that the ability to produce a consistent argument, relating evidence and explanation, increases with age with only a minority producing a consistent argument at age 9, rising to more than half at age 16.

In their study of investigative work, Millar *et al.* (1994) noted four kinds of response: an engagement frame, in which pupils engaged in activities without obvious plan or purpose; a modelling frame, in which they tried to produce a desired effect; an engineering frame in which they tried to optimize the effect; and a scientific frame. Although the use of the scientific frame increased from ages 9 through 12 to 14, still only a minority were using the scientific frame at age 14. The author (Watson 1994) studied groups of pupils carrying out practical investigations in a normal classroom setting and explored the factors that influence the approach of groups of pupils to an investigation. The findings showed that some pupils were found to carry out the practical activity with little purpose (engagement frame), but it was possible to influence the way the groups worked by explicitly focusing on the relationships between the research question and the data being collected by providing thinking time within the lesson structure, and by focusing classroom talk on the rationale behind the investigative strategies being used rather than on what was being done.

Openness of investigations

Carrying out whole investigations can be seen as a process of making a series of decisions. This means that pupils must have sufficient autonomy to be able to make decisions for themselves (Qualter *et al.* 1990; Woolnough 1991). Johnstone (1997), however, warns against overloading students' short-term memory and recommends structuring lessons in such a way as to reduce demands on the short-memory. Pupils' decision-making can also be supported by providing activities designed to scaffold pupils' thinking. The difficulty of achieving an appropriate balance between providing structure to support pupils' thinking, and providing a structure that directs pupils to perform an investigation, in a pre-determined way is illustrated in Watson and Wood-Robinson (1998). Teachers were often unaware of the extent to which they were making most of the important decisions. Teachers sometimes frame an investigation very heavily, leaving little for pupils to decide.

One aspect of giving pupils autonomy over their own learning is whether they can use the autonomy effectively to achieve the desired learning objectives. Watson *et al.* (1998) compared teachers' aims for specific investigation lessons with what their pupils thought they learned. Over 50 per cent of the teachers' aims were procedural, such as proposing hypotheses or planning a fair test, compared with 20 per cent of pupil responses. The majority of pupil responses (74 per cent) were about learning content compared with 33 per cent of teachers' responses. The mismatch between teacher and pupil perceptions is striking. Pupils concentrate on more surface features of investigations: If they are given a task 'to investigate the factors affecting the rate of sugar dissolving in water', they tend to see the purpose of the lesson being to learn about dissolving. Similarly, when they mentioned procedures, half their responses were very specific (for example, to learn to operate a balance). It is more difficult for pupils to recognize the gradual learning of scientific processes such as developing better fair testing strategies or approaches to planning. The implication, therefore, is that a key task for teachers is to *communicate* effectively to students what the educational *aims* of the investigation are; to differentiate between the aim of answering the question being investigated and the educational aims of activity.

Providing structure in an open situation

Two approaches to providing support in an open situation are through the overall structure of the investigation lessons, and through the organization of the activities within the different sections of the lessons.

Watson and Wood-Robinson (1998) describe a lesson structure used to report 32 investigations carried out by teachers of students aged 7 to 14. The lesson structure can be simplified into three phases:

- a thinking phase before collecting evidence;
- the collection of evidence;
- a thinking phase after collecting the evidence.

The initial thinking phase includes a focus on relevant knowledge and setting the problem in context, allowing pupils to clarify and refine the problem, and then planning what they are going to do. Typically, teachers spent about a third of the investigation time on these stages. In the next stages the focus is more on practical activity: making measurements and observations, recording and describing results and presenting their results in different forms. The focus then shifts back to thinking about what the evidence means: interpreting the results, evaluating the quality of the evidence collected and students reflecting on their own learning. Even when teachers planned time for the later thinking phase, this was often eroded because other parts of the investigation took longer than expected. The result was that the latter stages of the investigation were often either done alone at home or not done at all. This was seen as a serious weakness. For, if one of the purposes of investigative work is to develop a better understanding of evidence and its role in science, it is essential that time is given to critically analysing and discussing the interpretation and meaning of the data that students collect. The teacher has an important role in focusing on significant aspects of pupils' investigations, in challenging them to defend the quality of their evidence and arguments, and in creating a classroom environment in which pupils become critical of their own and others' evidence.

As well as providing support through the structure of the lessons, support can be provided within different stages of the lessons. Two illustrative examples are given below.

One way of providing guidance for pupils in the first thinking phase is using a sequence of displayed questions (variously called planning sheets, prompt sheets or planning boards). Watson and Wood-Robinson (1998) found that such scaffolds are very common, with many teachers using them most or all of the time. The effectiveness of such scaffolds depends on how they are used. They can be used like a traditional worksheet, almost like a recipe, to take the students through the investigation stage by stage. Used in such a prescriptive way, there is little opportunity for pupils to learn to make decisions. Alternatively, and more efficiently, they can be used to display and discuss the main features of students' thinking and can be used to expose this thinking for discussion either within a group or with the teacher.

Another aid to planning is the use of a variables table (Jones *et al.* 1992; Watson and Fairbrother 1993). In this study pupils identified the key variables relevant to their investigation as headings to columns on a blank table. Having brainstormed all the *things* that could be changed in their investigation, they wrote the variable they want to find out about (the outcome or dependent variable) as the heading for the last column. All the other variables that might affect the outcome were used as headings for the other columns. They then chose one of these as the independent variable and chose different values for it. The remaining columns represented control variables. The teacher was then able to come to the group, peruse the table and engage the pupils in discussion about their thinking. For example, Table 4.1 reveals that the group had not considered the effect of temperature on the rate of growth and had not written down exactly what aspect of growth they would examine.

Table 4.1   Example of a pupil's table of variables

Water	Soil	Plants (grass seedlings)	Amount of light	Growth
20 ml a day	Bag of soil	2 cm	Dark cupboard	
20 ml a day	Bag of soil	2 cm	Natural light	
20 ml a day	Bag of soil	2 cm	40w bulb on 24 h	

These two devices are not merely scaffolds for pupils, they are tools to expose pupils' thinking and make it more accessible to discussion and development.

Closing thoughts

Practical work is one of the hallmarks of science, and many educators argue that a science education without practical work fails to reflect the true nature of scientific activity. This has led to a widespread acceptance in many countries of a strong emphasis on pupils doing practical work. There is now a need to examine carefully the purposes of different kinds of practical activity in order to select appropriate strategies for achieving different aims. Some clear messages emerge from the literature.

- Scientific skills and processes are embedded in the context of the particular scientific purposes that they serve. Theory and practice are interrelated.
- Practical work has in the past been seen, not surprisingly, as a practical activity – doing things. Research indicates that this is not enough. Practical

work needs essentially to be about thinking: that is about trying to understand the relations between evidence and theory and to stimulate and challenge pupils. This is particularly true in investigations and so it is important to provide time for discussion and to encourage pupils to make their ideas explicit.

- Explicit teaching of concepts of evidence is recommended. There is, however, a difficult balance to be struck between providing support which gives students enough freedom to make decisions for themselves, and imposing a structure in which the teacher has made most of the important decisions.
- Autonomous learners need to be aware of the educational purposes of activities in which they are engaged (see Chapter 2 on formative assessment).

References

APU (1982) *Science in Schools: Age 13*, Research report. London: DES.

APU (1985) *Science at Age 15*, Report No. 1. London: DES.

APU (1988) *Science at Age 13*, Review report. London: DES.

Beatty, J.W. and Woolnough, B.E. (1982a) Why do practical work in 11–13 science? *School Science Review*, 63: 758–70.

Beatty, J.W. and Woolnough, B.E. (1982b) Practical work in 11–13 science: the context, type and aims of current practice, *British Education Research Journal*, 8: 23–30.

Driver, R., Leach, J., Millar, R. and Scott, P. (1996) *Young People's Images of Science*. Buckingham: Open University Press.

Foulds, K., Gott, R. and Duggan, S. (1997) *Science Investigations 1*. London: Collins.

Foulds, K., Gott, R. and Duggan, S. (1998) *Science Investigations 2*. London: Collins.

Gott, R. and Duggan, S. (1995) *Investigative Work in the Science Curriculum*. Buckingham: Open University Press.

Gott, R. and Murphy, P. (1987) *Assessing Investigations at Ages 13 and 15, APU Science Report for Teachers: 9*. London: HMSO.

Harmon, M., Smith, T.A., Martin, M.O. *et al.* (1997) *Performance Assessment in IEA's Third International Mathematics and Science Study* (TIMSS). Boston: Center for the Study of Testing, Evaluation and Educational Policy.

Head, J. (1982) What can psychology contribute to science education? *School Science Review*, 63(225): 631–42.

*Hodson, D. (1993) Re-thinking old ways: towards a more critical approach to practical work in school science, *Studies in Science Education*, 22: 85–142.

Johnstone, A. (1997) Know how your pupils learn and teach them accordingly, in D.L. Thompson (ed.) *Science Education and the 21st Century*. Aldershot: Arena Publishing.

*Jones, A., Simon, S., Black, P.J., Fairbrother, R.W. and Watson, J.R. (1992) *Open Work in Science: Development of Investigations in Schools*. Hatfield: Association for Science Education.

Kempa, R.F. and Dias, M.M. (1990) Students' motivational traits and preferences for different instructional modes in science education, *International Journal of Science Education*, 12: 195–203 and 205–16.

Kerr, J.F. (1964) *Practical Work in School Science*. Leicester: Leicester University Press.

Kuhn, D. (1989) Children and adults as intuitive scientists, *Psychological Review*, 96(4): 674–89.

Lubben, F. and Millar, R. (1996) Children's ideas about the reliability of experimental data, *International Journal of Science Education*, 18(8): 955–68.

Lynch, P.P. and Ndyetabura, V.L. (1984) Student attitudes to school practical work in Tasmanian schools, *Australian Science Teachers Journal*, 29: 25–9.

McRobbie, C.J., Roth, W.-M. and Lucas, K.B. (1997) Multiple learning environments in a physics classroom, *International Journal of Educational Research*, 27: 333–42.

Mahoney, M.J. (1979) Psychology of the scientist, *Social Studies in Science*, 9: 349–75.

*Millar, R. and Driver, R. (1987) Beyond process, *Studies in Science Education*, 14: 33–62.

Millar, R., Lubben, F., Gott, R. and Duggan, S. (1994) Investigating the school science laboratory: conceptual and procedural knowledge and their influence on performance, *Research Papers in Education*, 9(2): 207–48.

Qualter, A., Strang, J., Swatton, P. and Taylor, R. (1990) *Exploration: A Way of Learning Science*. Oxford: Blackwell.

Roth, W.M., McRobbie, C.J., Lucas, K.B. and Boutonne, S. (1997) The local production of order in traditional science laboratories: a phenomenological analysis, *Learning and Instruction*, 7: 107–36.

Scott, P. and Leach, J. (1998) Learning science concepts in the secondary science classroom, in M. Ratcliffe (ed.) *ASE Guide to Secondary Science Education*. Cheltenham: Stanley Thornes.

Song, J. and Black, P.J. (1991) The effect of task contexts on pupils' performance in science process skills, *International Journal of Science Education*, 13(1): 49–58.

Song, J. and Black, P.J. (1992) The effect of concept requirements and task contexts on pupils' performance in control of variables, *International Journal of Science Education*, 14(1): 83–93.

Strang, J., Daniels, S. and Bell, J. (1991) *Planning and Carrying out Investigations: Assessment Matters No. 6*. London: School Examination and Assessment Council.

*Swain, J., Monk, M. and Johnson, S. (1998) A comparative historical review of attitudes to the aims of practical work in science education in England: 1962, 1979 and 1997. Unpublished research paper, King's College London.

*Watson, J.R. (1994) Students' engagement in practical problem-solving: a case-study, *International Journal of Science Education*, 16(1): 27–43.

Watson, J.R. and Fairbrother, R.W. (1993) Open-ended work in Science (OPENS) Project: managing investigations in the laboratory, *School Science Review*, 75(271): 31–8.

Watson, J.R. and Prieto, T. (1994) Secondary science in England and Spain, *Education in Chemistry*, 31(2): 40–1.

Watson, J.R. and Wood-Robinson, V. (1998) Learning to investigate, in M. Ratcliffe (ed.) *ASE Guide to Secondary Science Education*. Cheltenham: Stanley Thornes.

Watson, J.R., Goldsworthy, A. and Wood-Robinson, V. (1998) Getting AKSIS to investigations, *Education in Science*, 177: 20–1.

*Watson, J.R., Goldsworthy, A. and Wood-Robinson, V. (1999a) One hundred and twenty hours of practical science investigations: a report of teachers' work with pupils aged 7 to 14, in K. Nielsen and A.C. Paulsen (eds) *Practical Work in Science Education: The Face of Science in Schools*. Copenhagen: Royal Danish School of Educational Studies.

Watson, J.R., Goldsworthy, A. and Wood-Robinson, V. (1999b) What is not fair with investigations, *School Science Review*, 80(292): 101–6.

Watson, J.R., Prieto, T. and Dillon, J. (1995) The effect of practical work on students' understanding of combustion, *Journal of Research in Science Teaching*, 32(5): 487–502.

White, R. (1988) *Learning Science*. Oxford: Blackwell.

Woolnough, B. (1991) *Practical Science*. Milton Keynes: Open University Press.

Woolnough, B. and Allsop, T. (1985) *Practical Work in Science*. Cambridge: Cambridge University Press.

5 The nature of scientific knowledge

Martin Monk and Justin Dillon

It is not difficult to think of activities that convey alternative messages about the nature of science. Note-taking from the board – a closed and directed activity – can carry a strong message of science as a pre-established body of knowledge that the student must master. With such an activity, science is presented as something that was done in the past, by people who are celebrated for their genius, and to which the student may contribute in the distant future after long and patient study. Presented in this way, in its final form, the student is confronted with an image of science as a set of facts.

Contrastingly, lots of genuinely open-ended investigational activities convey the message of science being contemporary, open to those who wish to have-a-go, and generating information to which students can add their contribution. It is the hands-on doing of science that is made problematic. The image of science is one of isolating variables and of monitoring relationships. The image of the scientist is one of the skilled experimentalist. The image of scientific knowledge is one of empirical accounts of the properties of the world that are available to those with the right skills.

Here is a third, perhaps less familiar, scenario. Students can be directed to some natural phenomena and asked to describe their experience. Their attention can be specifically directed to addressing the, 'What is happening?' question. The teacher can be particularly concerned about the careful articulation of ideas. Students can be encouraged to discuss in groups and report orally, or with different media. With such discussion, science is presented as a problem in conceptualization and persuasion. It is the argument that one uses to give a coherent account of what is happening, and why it happens, that is made problematic.

Science has features found in all three of the descriptions above. Scientists do work with empirical data. They may have to be good experimentalists to collect reliable data. Other scientists may spend their time inventing ideas to explain the phenomena in question. Some scientists may abandon the laboratory and spend most of their time codifying, simplifying and publicizing ideas. Scientists work within an historical tradition, and the scholarship of patient study of past ideas does help the advance of science. But, because doing science is not a unitary act – it is a constellation of different actions – a science education that focuses on one aspect, at the expense of others, can create in the minds of students a distorted image of science and the work of scientists.

What is in the chapter

The first section of the chapter looks at changing perspectives on the nature of scientific ideas and relates these to different forms of classroom practice. Philosophers, sociologists and historians of science have been interested in the status of scientific knowledge and its similarities and differences with other forms of knowledge for some time. The focus of their interest is on how scientific knowledge is generated and on how universal those methods might be. Understanding some of the implications of their work, classroom discussions about such issues can be used to show that the nature of scientific knowledge is problematic.

The second section of the chapter deals with the nature of scientific ideas as specified in *Science in the National Curriculum* (DfE 1995)and looks at some of the ways in which the documentation can be read. What does the documentation insist teachers must do? What latitude do teachers have? Although the documentation is UK-based the issues that emerge are general and could apply to science education in any country.

Lastly, there is a section in the chapter that looks at research evidence on views of the nature of science held by students and teachers. The question of whether teachers' ideas do, or do not, influence classroom practice is raised, as is the question of whether science teachers are constrained by other factors. This section approaches the nature of scientific knowledge in the classroom from the direction of what research shows is, or has recently been, happening, rather than what is possible in principle, or is required by law.

Changing perspectives on the nature of scientific knowledge

Questions about the status of scientific knowledge and the nature of scientific method are as old as discussions about natural phenomena. Different positions have been taken on (a) what is an adequate description of scientific method, (b) what status can one give to scientific knowledge and (c) how, if at all, does one distinguish science from non-science? (Chalmers

1982; Losee 1993). The discussion that follows is a very condensed version of work that has taken many brilliant thinkers lifetimes of scholarship. Inevitably, in packaging their ideas, we may have bruised the fruits of their labours.

Empiricism

As the nineteenth century turned into the twentieth, one popular and fashionable idea was for knowledge to be termed scientific only if it was derived empirically. A more severe form of this idea limits claims of scientific knowledge to observable quantities. Such a philosophy of science focuses attention on the laws of natural phenomena. This view is usually associated with the ideas of Ernst Mach (1943) and on what became known as the Vienna Circle. The philosophy of science of members of the Vienna Circle is often referred to as empiricism. Empiricists would claim that anything that is not expressed in terms of law-like behaviour, derived from empirical enquiry, is simply not science and not worthy of the name of science. A pre-occupation with empirical enquiry, together with the exclusion of any concerns over mechanisms or models, changes the empirical to empiricism. Empiricism concerns itself with questions over, 'What is happening?' and eschews questions of mechanism raised by, 'Why does this happen?' For the empiricist, the laws of observable behaviour of the natural world are the only things that can be given the status of scientific knowledge. For the hard-line empiricist, any suggestion of a mechanism or a model, which involves unobservable quantities, is ruled out from being termed scientific. This echoes Newton's claim that he made no hypothesis about the mechanisms of gravity, he simply worked with the law-like behaviour of falling objects. Quite how much truth there is in the tale that Boltzmann drowned himself in the Adriatic in 1906, because he was disturbed by Ernst Mach's insistence that atoms were an unnecessary hypothesis will never be known. But, the story does show how brutal severe classical empiricism can be as a philosophy of science.

   We can probably see a version of empiricism at work in most of the investigations carried out in contemporary UK secondary science education. Where students devise fair tests and collect sets of data to determine empirical relationships, with little regard being paid to any theory or underlying mechanisms, they are realizing the empiricists' views on the nature of science. From a psychological perspective, there may possibly be some justification for presenting students with this viewpoint. Novices cannot be expected to act as fully functioning experts, they are struggling with the skills of measurement and data recording and cannot be expected to keep a critique of the underlying theory or mechanism in view at the same time (Solomon *et al.* 1992). But, psychological expediency has a cost that is paid for in terms of the nature of scientific ideas presented to the students, and in terms of the understanding of the nature of science they thereby develop. The emphasis on the isolation and control of variables, which features in investigational work, often sits uncomfortably with the science content (theories, models and metaphors) of biology, chemistry and physics. Science teachers find it difficult to make investigational work genuinely investigational when they try to shoe-horn it into the teaching of content within the time constraints of an over-burdened curriculum.

Logical positivism

Scholars who thought the narrow focus of empiricism too severe developed a variation that allowed them to bring theories that incorporated unobserv-ables within their version of what constitutes scientific knowledge. Crudely, this next viewpoint admitted that scientists have theories about phenomena, and they use empirical tests to *confirm* those theories. This position became known as the positivistic viewpoint. Such a philosophy of science is known as positivism. The writings of Rudolf Carnap (1966) are often associated with this. To successfully incorporate unobservables into theories that are to be given the title 'scientific', positivists rely on the use of logical analysis and deduction. The umbrella, sheltering what constitutes scientific knowledge, is thereby extended and theories, models and ideas can be dignified with the label of scientific knowledge. Hence, logical positivism is a philosophy of sci-ence that suggests that scientific knowledge can be achieved through the combined operation of empirical study and logical deduction. An explicit value is that scientific knowledge can be *proved* to be 'true'.

What would positivism look like in the science classroom? The notion that one has a theory and then sets about an empirical enquiry to confirm it sits very comfortably with the teaching of content in many science classes. The sort of science education that has practical work developing out of theory, which, in turn, was developed from observation of phenomena, can approxi-mate to a science education underpinned by a logical positivist philosophy. Teachers are not engaged in the production of scientific knowledge. They are engaged in the reproduction of scientific knowledge. Teachers know, in advance of students' empirical enquiries, contemporary textbook ideas about the natural phenomenon under investigation. For the teachers, the activities that they plan for students in their schemes of work can therefore become illustrative rather than exploratory. The demonstration and class experiment takes on the mantle of a proof that the ideas about the natural phenomenon are 'true' (Millar 1998). Many science teachers work with such a tacit logical positivist outlook.

Although a science teacher may know the theory that is being illustrated, the students themselves may not be anywhere near so familiar with such theory as to be able to see how the evidence collected either supports the theory, or not. Thus, the students end up going through mechanical routines rather than testing theories that are derived from models which may incorp-orate unobservables. The work the students do is often neither very positivis-tic, nor are they able to provide any account that draws on logic. Derek Hodson (1990, 1992, 1993) has criticized practical work in science as being too much hands-on and not enough minds-on. However, as Solomon's work shows, all too clearly many students do come away from practical work with the idea that demonstrations and experiments *prove* scientific ideas to be true.

Falsificationism

Karl Popper (1972, 1980) is best known for holding an alternative position on the nature of science and for refuting positivism with his own idea which is popularly known as *falsificationism*. Falsificationists point out that the logic of the logical positivists should have alerted them to the impossibility of using empirical tests to prove a theory as there never can be sufficient evidence. Popper suggests that what can be shown through empirical tests is whether a theory is false. The Popperian view of the method of science makes use of hypothetico-deductivism. This is because the method involves making hypotheses and then making logical deductions from the combination of hypothesis and empirical results. For Popper, the theories, propositions, empirical tests and test results are all part of objective knowledge when they are in the public domain. Objective knowledge is always incomplete and pro-visional in principle (because it is always open to further test and refutation). Such a position marks a complete break between falsificationism and the school of thought associated with empiricism and positivism. For neither the positivist nor the empiricist would wish to give the appellation 'scientific' to such knowledge. The empiricist would wish to rule out the theories and mechanisms, while the positivist would be most unhappy with the provi-sional and tentative status.

Where can we see Popperian falsificationism in the science classroom? We would suggest that generally one cannot. We think students of science are hardly ever engaged in trying to disprove some theory or causal mechanism. Instead, they are more usually asked to carry out carefully controlled experi-ments which provide rhetorical support for the scientific view being advanced by their science teacher (Millar 1989; Ogborn *et al.* 1996). The social locations of the production and reproduction of scientific knowledge are different – research institute or school – and their social dynamics are dif-ferent. If Popperian falsificationism were to operate in science classrooms, then teachers would need to organize students' activities so that they could make their own conjectures, devise their own tests, and then take part in mini-conferences where ideas were critically evaluated. The student's reports on their investigations would need to be refereed by their peers before publication. Perhaps constructivism (see Chapter 3), as a viewpoint on learning, offers the opportunity for science teachers to re-conceptualize their teaching along these lines.

Science in a social context

Philosophizing over the nature of science went out of fashion from the 1960s onwards. The next generation of influential scholars turned to the history and practice of science in order to give an account of the nature of science. They focused less on prescription and, instead, tried to find patterns in description. The writings of Thomas Kuhn (1970) and Paul Feyerabend (1993) are often associated with a rise in popularity of this approach to the nature of scientific knowledge.

Thomas Kuhn is well known for suggesting that the historical record of scientific advance cannot be accounted for with a simple view of cumulative progression in scientific knowledge, but rather is better modelled with discontinuous changes in viewpoints, ideas, models and theories. Kuhn introduced the term 'paradigm' to describe such sets of viewpoints, ideas, models and theories that have some coherence. By way of example, we can see that Stahl's account of air and the process of burning (with his terminology of phlogisticated and dephlogisticated airs and calxes) forms a different paradigm to Lavoisier's accounts of the same phenomena (with the terminology of oxygen and oxides).

The writings of Paul Feyerabend are associated with the questioning of whether there really is a single scientific method whereby scientists can generate new scientific knowledge. Feyerabend pointed out that different methods are developed for different fields of scientific enquiry and, further, that paradigmatic shifts may develop new perspectives through using new methods. Feyerabend therefore suggested that the historical record of moves from one paradigm to another cannot be accounted for by rationality *alone*, because the logic of one paradigm will be contradicted by that of its successor. This led Feyerabend to the infamous idea that in scientific method, 'anything goes'.

In 1938 Reichenbach made the useful distinction between the context of discovery and the context of justification. The scientist, in working on a problem, and so bringing about a paradigmatic revolution, may use imaginative ideas and unconventional techniques with great tenacity to achieve previously unthought of conclusions. The context of discovery allows such variety. However, the discoverer then faces the burden of justifying the claim that the new viewpoint is more useful than the old. The context is now that of justification rather than of discovery. Evidence and logic, combined with communication skills, rhetoric and personal connections, are more important to the success of this enterprise than to the original discovery.

Science teachers also operate within the context of justification. Here, science education is different from science in its context of discovery. For science teachers generally expect to change students' ideas by rational means in the reproduction of knowledge in the school. They are given a greatly collapsed timescale in which to do this – years and not centuries. They work with people for whom science is just another course, and not a lifetime's passion. In this context, the positivism of school science stands in stark contrast to the falsificationism found within the research institute. Thus the single coherent canon of textbook science contrasts with the paradigmatic variety found in the history of scientific discovery. In schools, the training in the processes of science, and the censure of unacceptable or novel methods contrasts with the 'anything goes' of producing novel ideas about the natural world. Herein lies a paradox. Science education in its project to reproduce scientific knowledge is hard pressed to allow students of science to behave like fully fledged scientists. To do so, the students would need all the knowledge, skill, enthusiasm, passion, diligence, tenacity and time of the professional. This they do not have. They are students not practitioners.

Science in the National Curriculum

Science in the National Curriculum (SNC), which in England and Wales is the document that provides the national programme of study, is organized with material for each stage of the education system through which the students pass. Even though the specification and legal requirements are parochial to England and Wales, a discussion of SNC can illuminate general aspects of how curricular choices influence the nature of science presented in class-room activities. In SNC the statutory requirements on what must be taught about the nature of scientific ideas appear in the general preamble at each key stage. The specification on the nature of scientific ideas for grades 7 to 9 (11- to 14-year-olds) is that:

> Pupils should be given opportunities to:
> (a) consider the importance of evidence and creative thought in the development of scientific theories;
> (b) consider how scientific knowledge and understanding needs to be supported by empirical evidence;
> (c) relate social and historical contexts to scientific ideas by studying how at least one scientific idea has changed over time.
>
> (DfE 1995: 14)

The notion that claims about the properties of the world must be justified by evidence is very strong. It appears in both (a) and (b). There is reference to creative thought in (a) but, despite the fact that it is given mention, there is no indication of whether creative thought follows, precedes or is intermingled with the collection of supporting evidence.

Item (c), which gives the statutory requirement to study how at least one scientific idea has changed over time, offers no advice on the direction that study may take and the mechanisms of change that might be hypothesized. The fact that (a) and (b) have emphasized evidence could lead science teachers to interpret (c) from the point of view of a historical context that pays attention to changing evidence: when we have new evidence, we develop new scientific ideas. As noted above, the notion that evidence *alone* changes scientists' minds has been disputed by both Kuhn and Feyerabend. One could take the view that the engine of historical change is the way meanings are negotiated in the social institutions of science – research institutes, university departments, and professional societies – that are part of social life (Bloor 1993). This places the engine of change beyond the evidence itself into the realm of human agency.

We can see the requirements are open to different interpretations and this thereby allows teachers some flexibility in what they might do. If a teacher already has clear views on the nature of science and knows which activities can be used, this openness to interpretation might be a good thing. If you, the science teacher, are short of ideas (as is indicated by research reported in the last section of this chapter), the British Society for the History of Science can provide access to references for suitable materials (www.man.ac.uk/ Science_Engineering/CHSTM/bshs/).

The notes at the start of the general requirements, at the beginning of each key stage, indicate that work on the nature of scientific ideas should be applied across the science curriculum in both Experimental and Investigative Science and in the subject content areas. In seeking further guidance on the nature of scientific ideas, perhaps Experimental and Investigative Science has a stronger claim for attention, due to the earlier emphasis on evidence. The sections that are nested within Experimental and Investigative Science are:

- planning experimental procedures;
- obtaining evidence;
- analysing evidence and drawing conclusions;
- considering the strength of evidence.

The picture painted here is of science as an entirely experimental activity. The thrust is very much on getting experimental results and finding patterns in data. There is no hint that one might have a problem in deciding which data to collect. There is no suggestion that experimental design might be guided by some model of the phenomena under study, and the notion that this is negotiable (Latour and Woolgar 1986) is absent. In turning to Experimental and Investigative Science for help on the nature of scientific ideas, a particular view of the nature of science is made available: science as an unproblematic data-collecting activity.

What is rendered problematic is found in the last item on the list above – considering the strength of evidence – where the following are listed:

(a) to consider whether the evidence is sufficient to enable firm conclusions to be drawn;
(b) to consider anomalies in observations or measurements and explain them where possible;
(c) to consider improvements to the methods that have been used.

(DfE 1995: 16)

These concerns turn back on the empirical enquiry itself. They do not reflect the view that there may be something amiss with an enquiry other than insufficient, inadequate, or methodological problems with data collection. Somewhat shockingly, there is no mention of causality nor any obligation to question the underlying theoretical models, which are used to interpret the observations and measurements. The emphasis remains very much on evidence and its collection in the modern classroom, and the 1995 SNC does little to encourage anything else.

If one reads SNC at Key Stage 3 (KS3) in this way, then the nature of scientific ideas that emerges is of scientific knowledge consisting of relationships between self-evident variables that are related in regular law-like ways. The patterns of relationship are derived from empirical evidence. If one's view is that the nature of science and scientific knowledge consists of *only* this, and this alone, then it looks like a clear case of empiricism.

The parts of the curriculum that take the nature of science beyond empiricism are found in the content areas: Life Processes and Living Things; Materials and their Properties; and Physical Processes. However, although most science teachers want to deal with current scientific views on natural phenomena, if they rehearse these, and *only* do this, then science is presented as a set of beliefs based on recourse to authority and is no different from any other claims that are made on the basis of belief alone. Horton (1971) made this point when he compared the teaching of science to the propagation of African traditional thought.

Should the teacher be able to provide empirical evidence for the current scientific position, then science is presented as beliefs justified by uncontested evidence. Adding a bolt-on extra of asking students how to improve the experimental design, as suggested in the discussion on 'considering the strength of evidence' above, does not change the image of the nature of scientific ideas very much. It just suggests that the problem is getting the right experimental design (Braddick 1966). It does not suggest the problem is getting the right ideas, model or mechanisms for the phenomena in the first place (Hesse 1974). Without making the ideas, model or mechanism problematic, the scientist is still presented as just a collector and correlator.

If the teacher can add activities that help the students to (a) suggest, (b) adjudicate, between different ideas and models and mechanisms, then the nature of science presented to the students is much more one of conjecture and refutation: different models might be subjected to empirical test and refuted, or retained, for further refined tests. Here, human agents and their ideas enter the picture of the nature of science. To replicate this in science education, students need to be set a task where they can be given the opportunity to consider how different members of their class have different ideas about the same phenomenon (Driver *et al.* 1994).

Should the teacher have materials to hand that allow the students to look at how people have modelled the phenomenon in the past, then the image of science might be even more strongly focused on human activity. Human agency might enter the picture to a sufficient degree to counter-balance the notion that the progress of development of scientific knowledge is simply a matter of unfolding a logical puzzle. Students may be persuaded that science is more a matter of scientists struggling with ideas that are unclear and poorly articulated, and that it is only with hindsight, that we can see the logical necessity or fatal flaw in those ideas (Monk and Osborne 1997).

Research evidence on views of the nature of science

Lederman (1992: 332) identifies four lines of research into topics related to the teaching of the nature of science:

1 attempts to assess student conceptions of the nature of science;
2 curriculum innovation designed to 'improve' students' conceptions of the nature of science:
3 the assessment of, and attempts to improve, teachers' conceptions of the nature of science; and
4 identification of the relationship between teachers' conceptions, classroom practice, and students' conceptions.

In this last section of the chapter we will be looking predominantly at the first and the third of these areas.

Students' views, their sources and changes

Lederman and O'Malley (1990) report a survey of changes to 55 US high school grade 9 to 12 students' views of science after a year of science classes. The instrument used to monitor the students views consisted of items to which the students were invited to give open responses.

1 After scientists have developed a theory (for example atomic theory) does the theory ever change? If you believe theories do change, explain why we bother to learn about theories. Defend your answer with examples.
2 What does an atom look like? How do scientists know an atom looks like what you have described or drawn?
3 Is there a difference between a scientific theory and a scientific law? Give an example to illustrate your answer.
4 Some astrophysicists believe that the universe is expanding, while others believe that the universe is shrinking: still others believe that the universe is in a static state without any expansion or shrinkage. How are these different conclusions possible if all these scientists are looking at the same experiment and data?
(Lederman and O'Malley 1990: 227)

Lederman and O'Malley coded the students' responses as either absolutist or tentative, which roughly translate into the more orthodox terms of realist or instrumentalist. A realist view is associated with the notion that empirical enquiry enables scientists to describe and explain a reality that exists independently of the scientist's enquiry. For the realist, the scientist's enquiry uncovers or discloses reality. An instrumentalist view is associated with the notion that scientific ideas are tentative and can be arrived at through various methods and are valued because they work rather than because they are true. For instrumentalists, the descriptions and explanations produced are not evaluated with respect to their match to reality, but rather with respect to their use.

Over the year of the study, the students' responses to item 1 showed a shift away from absolutist views towards tentative views. Responses to item 2 showed the most marked shift towards tentative views. Responses to item 3 showed little change, while responses to item 4 showed most bafflement on the part of students, with the highest number of no responses and unclear responses. It was a naturalistic study and so the three teachers had not been asked to diverge from their normal teaching over the year. The authors conclude that these students' views on the nature of science developed out of the science to which they were exposed, and that the more science they learned, the less absolutist they became. A further conclusion might be that students' views of the nature of science are unlikely to be coherent nor can the views of a group be given a single characterization.

Zeidler and Lederman (1989) report a different survey of 409 US students, who studied with 18 high school biology teachers. The students completed an inventory at the beginning and end of a Fall semester. The students' responses to the inventory were categorized as showing either a realist view, or an instrumentalist view, of science. Shifts in the students' responses between the beginning and the end of the Fall semester were computed. Some students became more realist in their views while others became more instrumentalist. During that semester a researcher collected data on the classroom behaviour of the 18 teachers. Transcripts were made of classroom talk, observation schedules were used to record events, and copies of notes on blackboards were taken down. The teacher's classroom behaviour was then matched against the shifts in the student's responses. The researchers satisfied themselves that there was a link between the shifts in scores shown by the students and the classroom events in which they had taken part. From the evidence they collected, Zeidler and Lederman conclude that 'Teachers' ordinary language in the presentation of subject matter was found to have significant impact on students' conceptions of the nature of science' (Zeidler and Lederman 1989: 771).

Science teachers' views, choices and actions

There have been several attempts to survey teachers' views of the nature of science (Mellado 1998). The majority of the methods devised have been aimed at pre-service teachers in the USA. Various strategies have then been adopted to help pre-service teachers develop their own understanding of the nature of science. Nott and Wellington have gone so far as to produce a questionnaire that enables respondents to determine their 'Nature of Science Profile' (Nott and Wellington 1993), although this was developed mainly to provide a stimulus for discussion and thought rather than a precise psychometric test.

Koulaidis and Ogborn (1989) surveyed the views of 54 teachers and 40 student teachers associated with the Institute of Education, London, during 1984–85. They designed a questionnaire to monitor ideas on the nature of scientific method, the criteria of demarcation of science from non-science, ideas on patterns of scientific change, and ideas on the status of scientific knowledge. The teachers and student teachers who took part in the study were presented with statements like:

'As science changes or develops, new knowledge generally replaces ignorance or lack of knowledge.'
'New scientific knowledge follows no pattern of growth, being purely the result of what scientists happen to have done.'
'In general, the better of two competing theories is the one nearer the truth.'
'In general, the better of two competing theories is the one which gives more useful results.'

The teachers were invited to agree or disagree with the statements. Having analysed the teachers' responses, Koulaidis and Ogborn present a set of three broad tendencies that mark out constellations of views which characterize most of the teachers and student teachers in their sample. The picture is certainly not one of homogeneity across the group.

The subject area to which science teachers have a primary allegiance appears to have some influence on what one thinks of the nature of science. As well as differences being reported between the subject groups, the study also shows differences between student teachers and established teachers. Three-quarters of the biologists expressed relativist views. The chemists were the most eclectic. The physicists showed the strongest tendency to be rationalists. However, Koulaidis and Ogborn refer to these physicists as undecided rationalists due to their lack of consistency. In the Koulaidis and Ogborn study, student teachers were found to have somewhat different views on the nature of science than experienced teachers. This evidence suggests that not only does the subject content influence ones views, but it also looks as though experience in the classroom may modify those views.

These findings are quite a surprise, and contrast with other surveys, which have shown teachers and students to hold views that Koulaidis and Ogborn associate with empiricism and inductivism, for example, the Blanco and Niaz (1997) study carried out in Venezuela.

Lederman and Zeidler (1987) looked at the views on the nature of science and classroom actions of 18 American high school biology teachers, each with a minimum of five years' service (average 15.8 years' service). This is different from their 1987 study, where they were looking at classroom behaviour of teachers and students' ideas. They also carried out classroom observations of the teachers at work as well as giving the teachers a 48-item questionnaire to complete. From their analysis of the data, they concluded that the views the teachers expressed on the nature of science and scientific knowledge had little relationship, and therefore effect, on the actual classroom actions of the teachers.

For beliefs to have any effect on actions, there must be choices of alternative actions. For most science teachers, the choices are not formulated in terms of different approaches to the nature of science and scientific knowledge. Instead, the choices they face are managerial/technical. The influence of the imperatives of the classroom on teachers' actions is reported in a small-scale case study carried out by Tobin and McRobbie (1997). They looked in detail at the classroom practice of an experienced Australian chemistry teacher (called Mr Jacobs in the study) and his expressed beliefs about the nature of science. Although Mr Jacobs would talk in terms of science being an evolving discipline that was uncertain and changed over time, his classroom actions were not congruent with such a view. Instead, his actions were dominated by his own views on *the nature of learning* and the *students' needs*, gathered from long experience. His main goal was to help the students to pass the examinations and tests with good marks. To this end, his teaching methods reflected science as being a catalogue of facts that the students had to remember and repeat in examinations. Where students had to solve chemical problems, Mr Jacobs provided them with algorithms to follow. The students were entirely happy with Mr Jacobs's methods. This added to the conservative inertia of what Tobin and McRobbie call the enacted curriculum.

Is it any wonder that science graduates come to their teacher education courses convinced of the certainty of science and of the pre-eminent position of science among other school subjects? Or do they? The evidence from a range of studies seems to be that pre-service teachers' views, rather like children's views of science, can be classified as 'naive', 'inconsistent' and 'resistant to change'. Lederman *et al.* (1994: 142) conclude that 'it does not appear that pre-service teachers have well-formed knowledge structures . . . the structures that do exist are largely the result of college coursework and are often fragmented and disjointed with little evidence of coherent themes.' What we do not know much about is how pre-service teachers develop understanding and awareness of ways to teach the 'the nature of science'.

Anderson (1950), in a study of 56 Minnesota high school teachers, ascribed ignorance of knowledge of the scientific method to teachers being too busy imparting the factual aspects of the curriculum to be interested and/or concerned about how science works. It may be the case that they do not have an understanding of the link between what they teach, how they teach, and the impact on their students' view of science and scientists. It may also be the case that they do not believe that they can make much difference to their students or even that they should be making much difference!

Duschl and Wright (1989) found that issues of perceived students' needs, curriculum guide objectives and accountability all mitigated against consideration of the nature of science (see also Brickhouse 1989 and 1990). King's data (1991) indicated that teachers' lack of education in the history and philosophy of science left them bereft of suitable ideas about how such topics could be taught in the classroom.

Final comments

In 1996 the UK Government proposed that there should be a National Curriculum for Initial Teacher Training (NCITT). In terms of the nature of science, the proposals stated: 'As part of all courses [of initial teacher training], trainees must demonstrate that they know and understand the nature of science . . .' (TTA 1998: 19).

The idea that there is one 'nature of science' is something that many researchers (see, for example, Lederman 1992; Alters 1997) would object to. Another implication – that, if teachers 'know and understand the nature of science', then they will incorporate elements of the nature of science in their lessons – is equally contested. Lederman, for example, opines that the 'assumption that teachers' conceptions will necessarily be reflected in teachers' planned or actual behaviours is not supported by "the literature and logic" (Lederman 1997, personal communication). A final key assumption is that, even if teachers can teach the nature of science, then children will learn it. To say that this is questionable, particularly given the saturation of everyday culture with images of scientists some way from reality, would be in grave danger of understating the case.

References

Alters, B.J. (1997) Whose nature of science? *Journal of Research in Science Teaching*, 34(1): 39–55.

Anderson, K.E. (1950) The teachers of science in a representative sampling of Minnesota schools, *Science Education*, 34(1): 57–66.

Blanco, R. and Niaz, M. (1997) Epistemological beliefs of students and teachers about the nature of science: from 'Baconian inductive ascent' to the irrelevance of scientific laws, *Instructional Science*, 25(3): 203–31.

Bloor, D. (1993) *Knowledge and Social Imagery*, 2nd edn. Chicago: University of Chicago Press.

Braddick, H.J.J. (1966) *The Physics of Experimental Method*. London: Chapman and Hall.

Brickhouse, N.W. (1989) The teaching of the philosophy of science in secondary classrooms: case studies of teachers' personal theories, *International Journal of Science Education*, 11(4): 437–49.

Brickhouse, N.W. (1990) Teachers' beliefs about the nature of science and their relationship to classroom practice, *Journal of Teacher Education*, 41(3): 53–62.

Carnap, R. (1966) *Philosophical Foundations of Physics: An Introduction to the Philosophy of Science*. London: Basic Books.

*Chalmers, A. (1982) *What Is This Thing Called Science?* 2nd edn. Milton Keynes: Open University Press.

DfE (1995) *Science in the National Curriculum*. London: HMSO.

Driver, R., Asoko, H., Leach, J., Mortimer, E. and Scott, P. (1994) Constructing scientific knowledge in the classroom, *Educational Researcher*, 23: 5–12.

Duschl, R.A. and Wright, E. (1989) A case study of high school teachers' decision making models for planning and teaching science, *Journal of Research in Science Teaching*, 26(6): 467–501.

*Feyerabend, P. (1993) *Against Method*, 3rd edn. London: Verso.

*Hesse, M.B. (1974) *Models and Analogies in Science*. London: Sheed and Ward.

Hodson, D. (1990) A critical look at practical work in science, *School Science Review*, 70(256): 33–40.

Hodson, D. (1992) Redefining and reorienting practical work in school science, *School Science Review*, 73(264): 65–78.

Hodson, D. (1993) Re-thinking old ways: towards a more critical approach to practical work in school science, *Studies in Science Education*, 22: 85–142.

Horton, R. (1971) African traditional thought and Western science, in M.F.D. Young (ed.) *Knowledge and Control: New Directions for the Sociology of Education*. London: Collier-Macmillan.

King, B.B. (1991) Beginning teachers' knowledge of and attitudes toward history and philosophy of science, *Science Education*, 75(1): 135–41.

Koulaidis, V. and Ogborn, J. (1989) Philosophy of science: an empirical study of teachers' views, *International Journal of Science Education*, 11(2): 173–84.

Koulaidis, V. and Ogborn, J. (1995) Science teachers' philosophical assumptions: how well do we understand them? *International Journal of Science Education*, 17(3): 273–83.

*Kuhn, T.S. (1970) *The Structure of Scientific Revolutions*, 2nd edn. Chicago: Chicago University Press.

*Latour, B. and Woolgar, S. (1986) *Laboratory Life: The Construction of Scientific Facts*. Princeton, NJ: Princeton University Press.

Lederman, N.G. (1992) Students' and teachers' conceptions of the nature of science: a review of the research, *Journal of Research in Science Teaching*, 29(4): 331–59.

Lederman, N.G. and O'Malley, M. (1990) Students' preceptions of tentativeness in science: development, use and sources of change, *Science Education*, 74: 225–39.

Lederman, N.G. and Zeidler, D.L. (1987) Science teachers' conceptions of the nature of science: do they really influence teaching behaviour? *Science Education*, 71: 721–34.

Lederman, N.G., Gess-Newsome, J. and Katz, M.S. (1994) The nature and development of preservice science teachers' conceptions of subject matter and pedagogy, *Journal of Research in Science Teaching*, 31(2): 129–46.

*Losee, J. (1993) *A Historical Introduction to the Philosophy of Science*. Oxford: Opus Books, Oxford University Press.

Mach, E. (1943) *Popular Scientific Lectures*. Translated by T. McCormack, 5th edn. La Salle, IL: Open Court.

Mellado, V. (1998) Preservice teachers' classroom practice and their conceptions of the nature of science, in B.J. Fraser and K.G. Tobin (eds) *International Handbook of Science Education*. Dordrecht, Netherlands: Kluwer.

Millar, R. (ed.) (1989) *Doing Science: Images of Science in Science Education*. London: Falmer Press.

Millar, R. (1998) From rhetoric to reality, in J.J. Wellington (ed.) *Practical Work in School Science*. London: Routledge.

Monk, M. and Osborne, J. (1997) Placing the history and philosophy of science on the curriculum: a model for the development of pedagogy, *Science Education*, 81: 405–24.

Nott, M. and Wellington, J. (1993) Your nature of science profile: an activity for science teachers, *School Science Review*, 75(270): 109–12.

Ogborn, J., Kress, G., Martins, I. and McGillicudy, K. (1996) *Explaining Science in the Classroom*. Buckingham: Open University Press.

Popper, K. (1972) *Objective Knowledge*. Oxford: Oxford University Press.

Popper, K. (1980) *The Logic of Scientific Discovery*, 10th edn. London: Hutchinson.

Reichenbach, H. (1938) *Experience and Prediction*. Chicago: University of Chicago.

Solomon, J., Duveen, J., Scott, L. and McCarthy, S. (1992) Teaching about the nature of science through history: action research in the classroom, *Journal of Research in Science Teaching*, 29: 409–21.

Teacher Training Agency (TTA) (1998) *Initial Teacher Training: National Curriculum*. London: Teacher Training Agency.

Tobin, K. and McRobbie, C.J. (1997) Beliefs about the nature of science and the enacted science curriculum, *Science and Education*, 6(4): 331–54.

Zeidler, D.L. and Lederman, N.G. (1989) The effects of teachers' language on students' conceptions of the nature of science, *Journal of Research in Science Teaching*, 26(9): 771–83.

6 The role of language in the learning and teaching of science

Carys Jones

> When children learn language, they are not simply engaging
> in one kind of learning among many; rather they are
> learning the foundation of learning itself. The distinctive
> characteristic of human learning is that it is a process of
> making meaning.
>
> (Halliday and Martin 1993: 93)

The importance of language in the learning and teaching of science cannot be under-estimated. It is important for students in developing their scientific knowledge, and for teachers in understanding their students' learning processes. But research has shown that the ways in which language is used in the classroom by teachers and students are complex and the effects, though considerable, are often highly subtle and not self-evident. Therefore, it is important to develop an understanding of *what* happens with language, *why* it happens and *how* it happens.

Language is a tool that is used for expressing information and ideas. A variety of linguistic and non-linguistic modes are used for communication: listening and talking; reading and writing; discussing and arguing; narrating and describing; using actions, images and symbols – all of which are ways of signalling meaning and what linguists term 'semiotics'. How we communicate depends on a range of contextual factors such as the situation, the resources of the participants, the interaction among them, the topic and the purpose of the communication. Each mode has its own distinctive type of discourse (Lemke 1998b).

Scientists attempts to communicate by using a highly specialized language – the language of science – that incorporates more than just words. For scientists draw on a multitude of signs and symbols (a multi-semiotic system) to communicate their ideas: these include graphs, charts, diagrams and mathematical symbols and equations, as well as natural language. Therefore, science teachers need to be aware of how these signs and symbols can be instrumental in helping students to develop scientific knowledge and understanding in the classroom. Students also need to develop the ability to communicate and use the discourse of science in the classroom to optimize their learning. Therefore, science teachers need to develop a critical and sensitive awareness to how language works in the classroom. This awareness needs to take account of the constant dynamic interaction between language and thought. This is crucial to developing students' ideas and interpreting their beliefs. Inevitably, problems arise through miscommunication and need to be anticipated.

This chapter explores the multi-faceted nature of language within the context of teaching and learning science in the classroom. It does so in the following sections:

- the classroom as a site of multiple discourses;
- talking past each other;
- the nature of scientific discourse;
- talking science;
- reading science;
- writing science;
- investigating science by exploring language.

Each section attempts to illustrate the importance for the teacher of attending to language and the importance of language for learning science.

The classroom as a site of multiple discourse

Students bring into the science classroom a great variety of commonsense views derived from their individual experiences of the world. They also bring their own linguistic resources and communicative repertoires developed from early childhood in a variety of socal settings. These contribute to the social context of the classroom (Lemke 1990), where the students' own discourse gradually becomes extended to incorporate scientific discourse. Scientific discourse comes about through a complex process of socialization that involves code-switching – using language for different purposes with different social determinants for what may, and may not, be said – developing a sharing of experiences and, thereby, leading to the development of scientific knowledge and understanding.

Most children quickly become adept at code-switching in the situations they encounter, although it seems that middle-class children are much better prepared to develop a formal use of language than are working-class children (Lemke 1990; Sheeran and Barnes 1991). Just how much code-switching takes place within the classroom is impossible to determine. Frequent shifts between talking about individual feelings or problems, describing and discussing scientific content, and the language of classroom management are just a few that occur.

In multi-lingual classrooms language issues are complicated further. EAL (English as an Additional Language) students may have developed sophisticated strategies for coping with the varieties of English they encounter. For them, scientific English may be yet another type of experience (Rosenthal 1995). Their code-switching is an additional layer to using different languages. Houlton quotes an account, by a Muslim girl in an east Midlands primary school, which illustrates the complex ways in which some children operate linguistically in order to become a part of the world, or rather worlds, they find themselves in: 'I speak English at school, Gujerati on my way home to my friends. I read books at the mosque in Urdu and I learn passages from the Koran in Arabic ... My mum speaks Marathi' (Houlton 1983: 6).

Thus the classroom situation may bring together a rich variety of linguistic repertoires. Within the classroom the science teacher's own way of talking interacts with those of their students to channel, and develop, the ability to engage in, and share, scientific discourse.

Talking past each other

There is considerable agreement that a gap exists between the students' worlds and the world of the science they are meant to learn about (Lemke 1990). Consequently, too often, there is a breakdown in communication in the classroom, which leads to considerable frustration. A good example of spoken discourse, where one experienced teacher failed to convey scientific understanding to a student, is provided by Klaassen and Lijnse (1996). They describe a physics lesson during which the students are introduced to static forces. Having watched a video about forces, the teacher leads a discussion that passes through seven improvised situations in order to explain force to the pupils. The discussion moves towards a dialogue between the teacher and one pupil, called Jane, who is having difficulty working with the scientific concept, but is happy to believe the teacher. The teacher persists in trying to explain but fails to alter Jane's understanding and, finally, admits failure.

In their analysis, Klaassen and Lijnse conclude that the problem centres around the two differing conceptions of force that are held by the participants. Jane's general notion seems to be that force is a power that is applied to making someone, or something, behave in a certain way as in, 'She forced him to hand over the money', 'He forced the door open', or even 'He used force to open the door' and so on. She firmly holds on to her own use of the word 'force', whereas the teacher constantly uses the word to communicate a *scientific* meaning. Sadly, this point is not mentioned by the teacher. The conversation becomes fruitless and tedious, despite the teacher drawing on a variety of innovative strategies. This experience could have both short-term and long-term negative effects on Jane's, and possibly other students', enjoyment of the subject. For instance, they might conclude that *force*, as a concept used in science, is hard because it is difficult to talk in a scientifically acceptable way using the word 'force'.

Scientific discourse, as Lemke explains, can privilege the expert and alienate students, thus nurturing in the latter a 'certain harmful mystique of science' (Lemke 1990) where a student's intelligence, abilities and motivation are undermined. Lemke suggests that science is often presented as being difficult in the classroom because it is perceived as having a certain authority to state absolute facts and objective truths. This is achieved through the use of language that highlights objectivity, in preference to the subjectivity of experience, thus generating a conflict between the specialized technical and scientific discourse of experts and the commonsense talk of people in general.

Clearly, one of the main goals of science education is to guard against such conflicts and alienation. Ways need to be found to express the excitement of science without the frustration that is inherent with a new language, new ways of speaking or the consequent mis-negotiation of meaning.

The nature of scientific discourse

The language of science is 'a purpose-designed tool' (O'Toole 1996) used 'in specific contexts to meet specific needs' (Martin and Veel 1998). Like all languages, it is a dynamic language and, as new discoveries are made, established meanings are challenged and new meanings emerge. Martin and Veel cite three influences that explain why scientific discourse continually expands and changes.

First there is the emergence of *new fields* of scientific activity – as with the field of genetics this century. Either previously established knowledge is challenged or new concepts and ideas are constructed. Secondly, *new sets of social relations* emerge for the users of scientific discourse, requiring the language to be modified for different audiences, which, in its turn, influences the language of science. Finally, there is the influence of *new modes of representing and (re)producing knowledge*. New technologies, such as statistical packages and computer-aided graphics, enable data and knowledge to be represented and communicated in new forms bringing about a transformation in the way in which scientists communicate information among themselves and the acceptable forms (semiotic codes) that they use (Kress and van Leeuwen 1996; Lemke 1998a). A large number of scientific enterprises, from epidemiology to nuclear physics, have been transformed by such tools.

In the conversations of scientists doing science, their discourse moves from what is initially commonsense through to the uncommonsense (Wolpert

1992; Cromer 1993) and on to, what become, accepted scientific ideas. Sutton (1992) describes how scientific ideas emerge from the first tentative claims of a researcher to a textbook of established public knowledge some years, or even centuries, later. He explains how words *acquire* accepted stable, and often transformed, meanings. Having started out as figurative, the meaning becomes literal. For instance, early attempts to grapple with language to communicate descriptions of phenomena in static electricity involved the use of *charge of electricity*. This term was borrowed from the notion of charging – as in filling up with some quantity, for example, gun powder in a musket, or beer in a tankard. So insulated bodies could be charged – filled up – with electricity in the same way. The *electricity* in the 'charge of electricity' is now often taken-as-read and we just talk about charge – the contextual clue being provided by the word 'electric' or 'electricity'.

Sutton also discusses how borrowings may be across languages. The French words for small pieces of chemical equipment such as *pipettes* and *burettes* were easily accepted by the international community in the late eighteenth and early nineteenth centuries because their preciseness could not be matched by Anglo-Saxon derivations.

He then goes on to suggest that two functional systems of language are needed to communicate and develop science itself:

- a labelling system, which carries the established knowledge, accepted as fact;
- an interpretive system, which allows open, tentative and uncertain speculation.

Too often, he claims, teachers and students only use language as a label, leading to the perception that science is solely established knowledge, therefore static and to be assimilated as fact. Consequently, Sutton argues for studies of science-in-the-making so that students can see how meanings *are made* and how scientific language is crucially dependent on metaphor. Such studies would help to develop a dynamic view of science – as something that opens up the possibility of understanding the world in new ways – and that new ways of viewing science require new ways of expressing scientific phenomena through the use of familiar metaphors. Thus, in the eighteenth century, the emergence of clocks and their wide availability led to the use of metaphors and mechanisms associated with clocks, much as metaphors associated with computers can dominate our thinking today.

There are two important and distinctive linguistic patterns of scientific discourse that also convey the notion of fixed, objective knowledge rather than language in flux. These are:

- the use of the passive verb form;
- and the use of nominalization (Lemke 1990; Halliday and Martin 1993).

The passive voice dispenses with the personal voice and is particularly suited to writing up the method of an experiment. For example: 'The stem of a leafy shoot was cut . . .' is more suitable than 'I cut the stem of a leafy shoot . . .' because 'The stem' rather than the 'I' conveys the meaning and the focus of concern more directly and therefore more effectively (Halliday and Martin 1993: 194).

Nominalization, where nouns and noun phrases are derived from verbs or adjectives, construe phenomena as if they were things. Verbs such as 'move', 'explain', absorb' and adjectives such as 'stable', 'fluorescent' take on a more abstract meaning as nouns such as: 'motion', 'explanation', 'absorption', 'stability' and 'fluorescence'. Sometimes noun phrases can be structurally quite complex and can blur scientific meaning, as in the phrase 'Glass crack growth rate', which is a nominalization of: 'How quickly cracks in glass grow' (Halliday 1998).

This process results in a higher degree of abstraction that is metaphoric (Halliday and Martin 1993). Metaphors are at the heart of science; thus we now accept and see the flow of electric charge *is* a current, light *is* a wave and so on. Lakoff and Johnson (1980) provide an illuminating and comprehensive discussion about the use of metaphors, in general, for the interested reader.

Communication through more than words

Science has to use a multi-semiotic system to convey its meaning effectively – for natural language is poor at representing the nature of the phenomena with which science is concerned. Natural language is very good at conveying binary opposites, for example black and white, but poor at conveying the many different shades of grey that lie between them. The popular saying that a picture is worth a thousand words particularly applies to the communication of scientific ideas. For instance, consider trying to express all the information that we know about the eye in words rather than in one good diagram. Thus several modes of communication are needed to convey meaning in the science classroom: visual-graphical representations, mathematical equations, charts, tables, photographs, actions and so on as well as natural language (Lemke 1990, 1998b; Halliday 1993). Even then, scientific descriptions of the natural world are greatly simplified versions of that world, no matter how complex the translation into signs. Taken one step further, diluting the scientific discourse itself 'necessarily involves diluting the science that is taught' (Halliday and Martin 1993).

Rather than seeing words as having a fixed meaning, it is useful to think of language as having a *meaning potential* (Halliday 1978) that leaves room for the individual to interpret meaning within a context. This point is important in understanding how children can develop their own meanings about science. The language used, for example, in diagrammatic, tabular or graphic form, needs to be carefully and systematically selected and practised so that students learn to read the *intended* meaning, and present information themselves using scientific language in an appropriate manner but in a way that can be clearly understood – for example, see Long (1991) and Barnett (1992).

Students build up their own scientific understandings through integrating the various items of knowledge they receive and reconstructing it in their own minds. Lemke (1998a, 1998b) claims that this process of individual reconstruction comes naturally and easily. But, as he also emphasizes, how students individually interpret the knowledge they receive depends on their prior knowledge and experience – in this sense, the *text* that is already established in their minds. Meaning is therefore arrived at intertextually (Lemke 1992) by combining the text *out there* with the text *inside the student's mind*, not as objective or isolated information. In other words, each student is a meaning-maker. Although, in principle, we can never be sure how something is understood by another person, we can obtain perspectives on the meaning that is being communicated, which enables checks and double-checks, so the range of alternative meanings developed can be narrowed considerably.

Talking science

In the classroom, science is reconstructed for the student through a variety of linguistic interactions (Christie 1998). Sometimes the views that students bring from their own communities are not given enough attention and respect. From their study of children's questions for their conceptual depth, Wray and Lewis (1997) suggest that children's *unscientific* discourse may be under-valued. Rather, teachers may prefer to avoid difficult questions and carefully control the classroom language. Drawing on extensive research in science classroom interaction, Lemke (1990) argues that there is a power conflict between the teacher as adult and the student as child. He found from his analyses that the most common communicative practice takes the form of a triadic dialogue: the teacher asks a question; a student answers the question; the teacher evaluates the answer. There the dialogue ends. This pattern is referred to as the Initiation – Response – Follow-up (IRF) (Mehan 1979) or Initiation – Response – Evaluation (IRE) (Sinclair and Coulthard 1975) sequence. Such patterns of talk can inhibit students from working out relationships and meanings for themselves (see Chapters 1 and 3). For instance, too much language activity focused on *labelling* can discourage the processes of thought. Thinking can then become unfocused because the stimulus to think disappears. IRF/IRE sequences can have an important role, which is both useful and interesting, but only in a modified form which Cazden (1988) and Wells (1993) examine in a variety of lessons.

The dominance of low-level IRF activities often presents science to students as if it is objective, outside their experience and not the study of what people have thought and said about nature. As Gallas (1995) writes: 'Science does not originate from distance and the objectification of the world of nature: it begins with wonder, imagination and awe.'

Lemke found that student-centred types of spoken interaction, such as student debates and open-ended discussions, are distinctly *under-used* by science teachers. He cites other examples where the science teacher's dominance prevails: interrupting students rather than vice versa, pacing the lesson, controlling the topic and what is relevant to it, marking importance or creating mysteries as a way of arresting the class's attention, and, notably, rushing the more difficult parts to get through the syllabus: 'Teachers tend to feel they have done a better job if they have covered more material in a lesson.' Wray and Lewis (1997) and Gallas (1995) have similar concerns.

Students need to be personally engaged while they are learning science. This can only happen if they are initially allowed to draw on their own domains of knowledge and experience so as to make the links that help them to further their scientific understanding. Their understanding is then enhanced because they are encouraged to use the language of science. Learning science, therefore, means learning to use the language of science through opportunities to practise the talking, reading and writing of science (Wilson and McMeniman 1992).

It is useful to think of teaching science as a process of 'inducting someone into new ways of seeing and *new ways of talking*' (Sutton 1992, emphasis added). Talking about science in the classroom is fundamental to the restructuring of knowledge. Talk stimulates thinking and reflecting, which, in turn, leads to the articulation of thought and the development of scientific thinking. New *insights* are communicated and gradually become used as new *outlooks*.

Lemke (1990) demonstrates how content knowledge can be communicated and jointly constructed by teachers and students by introducing the useful concept of the 'thematic pattern': 'The most essential element in learning to talk science is mastery of the thematic patterns of each science topic. These patterns of semantic relationships among scientific terms are highly standardised in each field of science' (p. 27).

For example, the theme 'force' in science has a pattern of meaning expressed through such concepts and expressions as:

- the force of gravity;
- exerting a force;
- a force acting on something;
- a counterforce;
- force causes a change in the direction or speed of motion of a body;
- resultant force.

These may differ from the alternative patterns of semantic relationships that students may already have, as happened with Jane in the example discussed by Klaassen and Lijnse (1996). Lemke illustrates how thematic patterns in science can be followed through when the same basic pattern is expressed and discussed in different ways, using links and contrasting thematic patterns. The students are encouraged to develop their understandings of the pattern through talking about it and become able to adopt the appropriate semantic connections. Lemke provides a useful analysis of a classroom discussion about light being converted into heat where the students have the opportunity to challenge the teacher's statements by asking questions, or expressing doubt, and explaining why by drawing on their own alternative beliefs. Thus they learn to construct their own scientific discourse. Lemke suggests bridging the gap between colloquial and scientific language by explicitly discussing formal scientific style and the contrast with informal language. The shift from focusing on content to focusing on language at appropriate junctures requires artful management for 'there is a great deal of repetition, use of examples, and implicit use of terms and of principles across a variety of contexts in good teaching' (1990: 24).

Sutton (1992) advises that *doubt* should be a key feature of learning. 'Puzzling and telling are complementary.' Sutton argues that 'science lessons should be the study of systems of meaning which human beings have built up': thus presenting an evolutionary view of science as something that is open to change, which is constructed and reconstructed as human beings gain new insights into the knowledge they already have and combine it with existing knowledge as new discoveries emerge.

Gallas (1995) argues that students, together with the teacher, need to move towards 'an inclusive kind of talk about science where everyone is admitted' (p. 3) to permit the appropriation of a science discourse. In her study Gallas too starts from the premise that children have their own theories, and need to be able to articulate them somehow in order to move forward in their thinking. To begin with, this means activating students' prior knowledge in order to establish, as far as possible, what students know, don't know and what misconceptions they might have (see Wray and Lewis 1997). Though, as she points out, this is by no means straightforward. Whereas theories 'presented to the world are rational and orderly', Gallas (1995) writes, 'the process of generating ideas and developing and changing theories is . . . complex and often disorderly' (p. 43) but 'incorrect theories are better than silence!' (p. 99).

Her account of her 'Science Talks' provides some useful insights into how children can be encouraged to control their own discussions with the minimum amount of teacher intervention. She emphasizes the importance of asking appropriate questions as long as they are selected carefully, and particularly highlights the first question – which must be a seminal question to trigger children's thinking and consequently their own questions:

> I began our first Science Talk with a focusing question, one that I still use every year with each new class because it illustrates the nature of the questions we will consider. 'Why do the leaves change colour?' . . . I am continuously astonished by the enthusiasm and originality of the children's theories and the eagerness with which they adopt the talks as their own.
>
> (Gallas 1995: 19)

She further observes how the children's thinking develops as they discuss with each other, co-constructing their knowledge, and how such discussions provide clues about the depth of their knowledge as well as about their beliefs of the world. Through analysing such discussions, Gallas also learned more about her own role as teacher, and about the ways she could keep in the background and intervene only at appropriate junctures to keep the discussions on course thematically.

Reading science

Reading as a part of the learning process is an area that is little understood or researched in science education. Too little attention has been given to developing children's scientific literacy (Davies and Greene 1984; Wray and Lewis 1997) through the use of reading, which is a highly under-used activity in science classrooms.

It is commonly observed that the genre (style of writing) in science texts can be much more difficult than that associated with subjects in the humanities, such as a story in narrative form in English, to which many students are highly receptive (Davies and Greene 1984). Martin suggests that either science textbooks are too difficult for students to read, without considerable guidance, or the opposite happens: 'an attempt has been made to make science more accessible by downplaying science literacy' (Halliday and Martin 1993: 202). Nor can we take it for granted that science textbooks will engage all students' interests equally (Eltinge and Roberts 1993; Jones 1997).

Davies and Greene (1984) warn of the danger of passive reading, by which they mean that students do not process anything. They suggest that passive reading comes about because the reading purposes are vague and unspecific, and the reading activity is solitary rather than *shared*. They argue that science texts have to be read for a specific purpose to engage students' interests and to promote their learning, and they describe in detail their study of a variety of scientific texts that typify school science. Their purpose was two-fold: to gain a deeper understanding of the key features of text types and to suggest ways in which students can be helped with their reading.

The first aspect of their analysis revealed that the quality of published texts is variable and they need to be critically evaluated if they are going to be used. Their analysis is particularly useful because it shows how science teachers can gain an enhanced awareness of text types as well as a critical overview of texts. It highlights the content and structure of the text – what is called the *frame*. They identified three broad categories: those that deal with activities, those that deal with phenomena, and those that deal with ideas. Within these, they found seven types of text representing different frames:

- instruction;
- classification;
- structure;

- mechanism;
- process;
- concept-principles;
- hypothesis-theory.

A further part of their analysis examined the language that is used for the functions that comprise these frames. For example, instruction texts have a frame that introduces '*actions*, or steps, or procedures, *materials, apparatus, cautions* or *conditions, results*, or *effects*, or *outcomes of procedures, interpretation of results*' (Davies and Greene 1984: 90; original emphasis).

The second aspect of their work explored the use of text for learning science. Whereas normal reading is receptive, reading in science has to be *reflective*, requiring the text to be re-read and considered. To help with learning this process, they suggest a variety of activities called DARTs (directed activities related to text): reconstruction activities, requiring text to be re-assembled such as the sequencing of scrambled segments of a text; completion activities such as filling in gaps with a word or a phrase, or completing a diagram or table; analysis activities such as text underlining, segmenting where pupils have to provide headings for sections of text, or diagram construction; and text extension where questions about the information are generated by the students. Their study provides seminal help to science teachers in developing an analytical approach to understanding the potential of texts and textbook teaching materials. Other useful work on classifying the content of science texts in terms of readability has been done by Barnett (1992), Perkins (1992) and Long (1991).

Wray and Lewis (1997), in their investigations of reading and writing at primary level in the UK, found several limiting practices that held back children's comprehension of non-fiction text: reading materials made available were narrow in range and quality; they were restricted to the narrative and not subject-specific; and children had difficulty in using library sources. As they argued, it is important to teach children to adopt a critical approach to literacy from the start. Children can become critical thinkers and researchers through guided reading for specific purposes (Wray and Lewis 1997). Wray and Lewis also studied effective strategies to help students find and retrieve information, and read interactively. They recommend *think-aloud* techniques, where the teacher shows why she/he refers to a particular resource to find information, discussing a variety of texts that present different viewpoints about the world, and encouraging children to produce their own texts.

Writing science

Science teachers often question how far students need to develop scientific literacy through transactional writing – the type of factual and impersonal writing that is used to convey information. Many children need a lot of practice to acquire it. Sheeran and Barnes (1991) discuss a variety of views about whether or not, and to what extent, students should be encouraged to learn to write in the standard genres of scientific discourse. On the one hand are the views that scientific writing trains the mind to think scientifically, in a disciplined and structured way; that it encourages students to gain access to the public domain of scientific knowledge, and, thereby, to become interested in science as a career; and that it is necessary for passing examinations. The counter-arguments are that children need to be able to express their thoughts freely in their own language; that focusing too much on language means neglecting the content; that factual writing will develop naturally; that scientific discourse alienates students; and that expressing science in words is difficult.

Students' writing can be very deceptive, particularly in cases where the writing is fluent and well structured, but the content is superficial (Wade and Wood 1979; Jones 1997). Expressive writing blurs the quality of the scientific thought and makes assessment even harder for the teacher. Sheeran and Barnes (1991) found in their studies that, even after discussing an experiment, students' written reports of it did not seem to reflect any real understanding of what had happened. Their evaluation of examples of students' writing about science revealed that students' work can be easily misinterpreted and wrongly assessed. However, they argue that students need to be able to write in a way that they feel comfortable with, as then they are more likely to display the extent of their scientific thinking and knowledge. As Cortazzi (1995) suggests, the genres of discourse need to be regarded as broad tendencies rather than as rigid norms. Thinking develops through being personally engaged, which may be helped by writing in a personal genre. However, at the same time, this type of expressive writing is undoubtedly restricted in its potential to develop a scientific mind because it involves a personal response, a feature normally excluded from scientific writing. It may also lead to inaccuracy (Halliday and Martin 1993: Chapter 9). Undoubtedly, writing is a complex activity. But it can fulfil both cognitive and metacognitive roles (Bereiter and Scardamalia 1987; Rowell 1997). It has a generative power because it brings thought to consciousness, and thus offers the opportunity to re-organize meaning. Therefore, it is an important activity in any subject.

Rowell (1997) suggests that we need to think of writing as a social practice with three overlapping dimensions: interpretive, knowledge-transforming and discursive.

- As an *interpretive activity*, writing about science in the ways that *they* understand it, helps students appropriate knowledge. This helps them to construct extensions of understanding by moving beyond the everyday through their own use of language.
- As a *knowledge-transforming activity*, students develop critical thinking skills such as scientific reasoning, making predictions, making inferences, assessing evidence, drawing conclusions and judging argumentation

(Baker and Saul 1994; Keys 1994). This can be highly empowering even if it is a very gradual process (Bereiter and Scardamalia 1987).

- As *discursive practice*, students develop an awareness of the need to negotiate their meaning in order to convey their own understandings.

If not nurtured properly, writing can have negative effects because it 'deprives language of the power to intuit' (Halliday 1993). In other words, if students are required to present their scientific ideas in unfamiliar genres, using set linguistic patterns where no thinking processes are involved, the language no longer conveys any meaning, neither for the reader nor for the writer (Hoban 1992). Such writing can *freeze* language, as Halliday puts it, and lead to a deterioration in performance, particularly in the case of less competent students (Bereiter and Scardamalia 1987).

Different kinds of writing activities lead students to focus on specific types of information and to think about that information in specific ways. For example, analytic writing focuses on selective parts of text; summary writing focuses on the text as a whole; and short answer questions focus on particular information. In each case, students are encouraged to think about the information holistically or analytically, with attention either to specific detail or to the whole picture (Langer and Applebee 1987).

Students of science do need to become aware of different genres of writing as representations of different types of knowledge. The genres of non-fiction writing are:

- instruction,
- explanation,
- argument and
- discussion.

They need to be developed separately. Wray and Lewis (1997) provide helpful outlines of such genres. The development of writing can be helped through making formal writing an occasional event and explaining it properly (Sutton 1996) or through collaborative work and with prompts from the teacher as necessary (Bereiter and Scardamalia 1987; Keys 1994; Rowell 1997). Wray and Lewis also advocate the use of *writing frames*, i.e. skeleton outlines that include the beginnings of appropriate phrases, and provide a form of scaffolding that is particularly helpful in prompting young children's learning. These are now used extensively in primary science.

Investigating science by exploring language

Science classrooms are places where meaning is constantly being negotiated, no matter how tacitly, no matter what the power relationships are, through a variety of semiotic modes. Christie (1998) shows how a topic, over a series of lessons, may start as something new, and unfamiliar, and gradually become an integral part of a student's experience. She highlights the usefulness of examining particular types of discourse in order to understand the way a topic can be developed systematically from beginning to end in the classroom.

For example, an investigation comprises several stages, each with its own discourse. The discussion that permeates the stages of the investigation itself needs to continue into its writing-up so that cognitive links can be made between the actual event, the informal talk about the event, and the formal recording of it. The first introductory stage might be an exploratory, informal discussion about how to test a hypothesis. A later stage might be to conduct an experiment according to a set of formally written instructions. At some point it will be necessary to write up the experiment in an acceptable scientific genre. Writing the report should be an integral part of a process during an investigation rather than the final activity, thus encouraging students to reflect on the experiences of the investigation, making it an experiential as well as a coherent event.

Each of the four parts of the traditional report – the aims, the methods, the results and the conclusions – has its own style because each represents one distinctive aspect of the experiment. Each part needs to be discussed separately as an event so that it can be recorded meaningfully. Initially, this alone may be a time-consuming process, since the teacher needs to direct the discussion towards a common consensus about the way it should be written up. Discussion needs to switch between talking about each *event* and talking about *recording the event*. With this groundwork, students will come to understand how to write up investigation reports through reconstructing each part with the benefit of hindsight and reflection. With time, students will come to manage this on their own, once they have understood that the writing-up process provides an opportunity to deepen their understanding of the investigation, and consolidate what has been learned. Such practice helps students to enhance their language awareness through thinking critically about how their own knowledge and understanding can be represented in appropriate written scientific language.

Conclusions

Raising the students' own language awareness at appropriate moments requires artful handling. The art is to know if, when, and how, language should be temporarily and consciously brought to the fore – that is to shift to talking about *how we are describing* the phenomenon rather than describing it itself – in order to enhance, rather than hinder, the students' growing understanding.

References

Baker, L. and Saul, W. (1994) Considering science and language arts connections: a study of teacher cognition, *Journal of Research in Science Teaching*, 31: 1023–37.

Barnett, J. (1992) Language in the subject classroom: some issues for science teachers, *Australian Science Teachers Journal*, 38(4): 8–13.

Bereiter, C. and Scardamalia, M. (1987) *The Psychology of Written Composition*. Hillsdale, NJ: Lawrence Erlbaum Associates.

Cazden, C.B. (1988) *Classroom Discourse. The Language of Teaching and Learning*. Portsmouth: Heinemann.

Christie, F. (1998) Science and apprenticeship: the pedagogic discourse, in J.R. Martin and R. Veel (eds) *Reading Science*. London: Routledge.

Cortazzi, M. (1995) Do we have to write about it now? in J. Moyles (ed.) *Beginning Teaching: Beginning Learning in Primary Education*. Buckingham: Open University Press.

Cromer, A. (1993) *Uncommon Sense: The Heretical Nature of Science*. New York: Oxford University Press.

*Davies, F. and Greene, T. (1984) *Reading for Learning in the Sciences*. Edinburgh: Oliver and Boyd.

Eltinge, E.M. and Roberts, C.W. (1993) Linguistic content analysis: a method to measure science as inquiry in textbooks, *Journal of Research in Science Teaching*, 30(1): 65–83.

*Gallas, K. (1995) *Talking Their Way into Science*. New York: Teachers College Press.

Halliday, M.A.K. (1978) *Language as Social Semiotic: The Social Interpretation of Language and Meaning*. London: Edward Arnold.

Halliday, M.A.K. (1993) Towards a language-based theory of learning, *Linguistics and Education*, 5: 93–116.

Halliday, M.A.K. (1998) Things and relations: regrammaticising experience as technical knowledge, in J.R. Martin and R. Veel (eds) *Reading Science*. London: Routledge.

Halliday, M.A.K. and Martin, J. (1993) *Writing Science*. London: Falmer Press.

Hoban, G. (1992) Teaching and writing in science: match or mismatch? *Australian Science Teachers Journal*, 38(4): 36–8.

Houlton, D. (1983) Responding to diversity, in Schools Council, *Teaching in a Culturally Diverse Society: Papers from a National Conference*. London: Information Centre, Schools Council.

Jones, C.L. (1997) Communicating through writing about processes in science, *Journal of Biological Education*, 31(1): 55–64.

Keys, C.W. (1994) The development of scientific reasoning skills in conjunction with collaborative writing assignment: an interpretive study of six ninth-grade students, *Journal of Research in Science Teaching*, 31(9): 1003–22.

*Klaassen, C.W.L.M. and Lijnse, P.L. (1996) Interpreting students' and teachers' discourse in science classes: an underestimated problem? *Journal of Research in Science Teaching*, 33(2): 115–34.

Kress, G. and van Leeuven, T. (1996) *Reading Images: The Grammar of Visual Design*. Geelong Victoria: Deakin University Press.

Lakoff, G. and Johnson, M. (1980) *Metaphors We Live By*. Chicago: University of Chicago Press.

Langer, J.A. and Applebee, A.N. (1987) *How Writing Shapes Thinking: A Study of Teaching and Learning*. Urbana, IL: National Council of Teachers.

*Lemke, J.L. (1990) *Talking Science: Language, Learning And Values*. Norwood, NJ: Ablex.

Lemke, J.L. (1992) Intertextuality and educational research, *Linguistics and Education*, 4: 257–67.

Lemke, J.L. (1998a) Multiplying meaning: visual and verbal semiotics in scientific text, in J.R. Martin and R. Veel (eds) *Reading Science*. London: Routledge.

Lemke, J.L. (1998b) *Teaching All the Languages of Science: Words, Symbols, Images and Actions* [World Wide Web Site]. Available: http://academic.brooklyn.cuny.edu/education/jlemke/papers/barcelon.htm.

Long, R.R. (1991) Readability for science, *School Science Review*, 73(262): 21–33.

Martin, J.R. and Veel, R. (eds) (1998) *Reading Science*. London: Routledge.

Mehan, H. (1979) *Learning Lessons: Social Organization in the Classroom*. Cambridge, MA: Harvard University Press.

O'Toole, M. (1996) Science, schools, children and books: exploring the classroom interface between science and language, *Studies in Science Education*, 28: 113–43.

Perkins, K. (1992) How will I know what I think until I see what I say? *Australian Science Teachers Journal*, 38(4): 20–7.

Robinson, S. (1992) Writing in science – a science teacher's perspective, *Australian Science Teachers Journal*, 38(4): 42–4.

Rosenthal, J.W. (1995) *Teaching Science to Language Minority Students*. Clevedon: Multilingual Matters.

*Rowell, P.M. (1997) Learning in school science: the promises and practices of writing, *Studies in Science Education*, 30: 19–56.

*Sheeran, Y. and Barnes, D. (1991) *School Writing*. Milton Keynes: Open University Press.

Sinclair, J. and Coulthard, M. (1975) *Towards an Analysis of Discourse*. Oxford: Oxford University Press.

*Sutton, C. (1992) *Words, Science and Learning*. Buckingham: Open University Press.

*Sutton, C. (1996) Beliefs about science and beliefs about language, *International Journal of Science Education*, 18(1): 1–18.

Veel, R. (1992) Engaging with scientific language: a functional approach to the language of school science, *Australian Science Teachers Journal*, 38(4) 31–5.

Veel, R. (1998) The greening of school science: ecogenesis in secondary classrooms, in J.R. Martin and R. Veel (eds) *Reading Science*. London: Routledge.

Wade, B. and Wood, A. (1979) Assessing writing in science, *Language for Learning*, 1(3):131–8.

Wells, G. (1993) Reevaluating the IRF sequence: a proposal for the articulation of theories of activity and discourse for teaching and learning in the classroom, *Linguistics and Education*, 5(1): 1–38.

Wilson, J. and McMeniman, M. (1992) The mediating role of language in effective science learning: teacher-in-action and student perceptions, *Australian Science Teachers Journal*, 38(4): 14–19.

Wolpert, L. (1992) *The Unnatural Nature of Science*. London: Faber and Faber.

*Wray, D. and Lewis, M. (1997) *Extending Literacy: Children Reading and Writing Nonfiction*. London: Routledge.

7 Students' attitudes towards science

Shirley Simon

During a recent journey on the London Underground my attention was drawn to a group of teenage girls who were having a heated discussion about the electrolysis of brine. I could hear them talking about 'sodium ions' and 'chlorine gas'. One girl had her science folder open upon her knees and took the role of chief explainer, while three others chipped in with challenging questions and alternative ideas. The girls were totally absorbed in trying to understand various terms, in what was happening at the anode and cathode, and in the movement of different ions. In short, for that ten minutes of their lives they were totally involved in the electrolysis of brine. The train was quite crowded and I soon noticed that other passengers had ceased talking and were, like me, listening to the girls' discussion. Most looked totally bemused. An elderly couple sitting opposite me were staring at the girls, fascinated. Is the scenario of teenage girls being stimulated by the electrolysis of brine so strange?

A well-established aim of school science is to promote enthusiasm for the subject, not only to encourage choice of science post-16 and subsequent careers in science, but also to enhance all students' interest in scientific issues in adult life. Sadly, evidence shows that, for many students, this aim is far from realized, for the experience of school science leaves many with the feeling that science is difficult and inaccessible. In recent years there has been a range of studies concerning students' attitudes to science, focusing on factors which influence attitudes and subject choice post-16. Though some factors are outside the influence of school, many are concerned with classroom practice. In this chapter I will review significant research which has taken place in this area, focusing on what is meant by attitudes to science, why they have been extensively researched, and what is known about such attitudes. I will then examine the major influences on attitudes to science and subject choice that have been identified in research, and draw some implications for classroom practice from the range of studies undertaken.

Significant research

What are attitudes?

Thirty years of research into this topic has been bedevilled by a lack of clarity about what attitudes to science are. An early contribution towards its elaboration was made by Klopfer (1971), who categorized a set of affective behaviours in science education as:

- the manifestation of favourable attitudes towards science and scientists;
- the acceptance of scientific enquiry as a way of thought;
- the adoption of 'scientific attitudes';
- the enjoyment of science learning experiences;
- the development of interests in science and science-related activities;
- the development of an interest in pursuing a career in science or science-related work.

Research into attitudes *towards* science is further complicated by the fact that attitudes do not consist of a single construct but rather of a large number of sub-constructs, all of which contribute in varying proportions to an individual's attitudes to science. Various studies have incorporated a range of components in their measures of attitudes towards science, including:

- the perception of the science teacher;
- anxiety towards science;
- the value of science;
- self-esteem at science;
- motivation towards science;
- enjoyment of science;
- attitudes of peers and friends towards science;
- attitudes of parents towards science;
- the nature of the classroom environment;
- achievement in science;
- fear of failure on course.

Ramsden (1998) draws on definitions of attitudes, which include cognitive, emotional and action tendency components – action tendency being that which leads to particular behavioural intents. For example, that of Shaw and Wright (1968), who suggest that 'attitude is best viewed as a set of affective reactions towards the attitude object, derived from concepts of beliefs that the individual has concerning the object, and predisposing the individual to behave in a certain manner towards the object' (Ramsden 1998: 13).

However, of themselves, attitudes may not necessarily be related to the behaviours a person actually exhibits (Potter and Wetherall 1987): for example, a pupil may express interest in science but avoid publicly demonstrating it among his or her peers, who regard such an expression of intellectual interest as not being the 'done thing'. In such a case, motivation to behave in a particular way may be stronger than the motivation associated with the expressed attitude, or alternatively, anticipated consequences of a behaviour may modify that behaviour so that it is inconsistent with the attitude held.

Consequently, it is behaviour rather than attitude that has become a focus of interest and which has led researchers to explore models developed from studies in social psychology, in particular, Ajzen and Fishbein's theory of reasoned action (1980) – a theory which is concerned fundamentally with predicting behaviour. This focuses on the distinction between attitudes towards some 'object', and attitudes towards some specific action to be performed towards that 'object', for example between attitudes *towards* science and attitudes towards *doing* school science. Ajzen and Fishbein argue that it is the latter kind of attitude that best predicts behaviour. Their theory represents a relationship between attitude, intention and behaviour. Behaviour is seen as determined by intention, and intention is a joint product of attitude towards the behaviour and the subjective norm (that is beliefs about how other people would regard one's performance of the behaviour).

The theory of reasoned action has been successfully applied to some attitude and behaviour studies in science education (for example Crawley and Coe 1990; Norwich and Duncan 1990; Crawley and Black 1992). For instance, Koballa (1988), Oliver and Simpson (1988) and Crawley and Coe (1990) have all found that social support from peers and attitude towards enrolling for a course are strong determinants of student choice to pursue science courses voluntarily, which suggests that the theory has at least some partial validity. The main value of such a theory is its help in determining salient beliefs which can then be reinforced or downplayed to affect relevant behavioural decisions by students, such as *girls don't do science*. Furthermore, this theory points towards the need to draw a demarcation between school science and science in society. It is the perception of school science, and the feelings towards undertaking a further course of study, which are most significant in determining children's decisions about whether to proceed with further study of science post-16.

Why research attitudes?

The purpose of much attitude research in science education has been to identify features of a 'problem' (Ramsden 1998) to do with the ways in which young people's experiences or perceptions of school science appear to alienate them from science. This alienation is of concern to teachers, as their job satisfaction is likely to be strongly influenced by their pupils' affective responses to what is on offer in science lessons, perhaps even more than by their cognitive responses. It also has important implications for the uptake of science post-16 and the pursuit of scientific careers.

Concerns about attitudes to science and the uptake of science post-16 are not new. Over 20 years ago, Ormerod and Duckworth (1975) began their review on the topic of pupils' attitudes to science with the following comment:

> In 1965 a thorough inquiry began into the flow of students of science and technology in higher education. The final report (Dainton 1968) laid particular emphasis on the phenomenon which had become known as the 'swing from science'. Several explanations were suggested for the swing, among them a lessening interest in science and a disaffection with science and technology amongst students.
>
> (Ormerod and Duckworth 1975:1)

Since this review, there have been other major reviews of research into attitudes to science (Gardner 1975; Munby 1983; Schibeci 1984), which make reference to over 200 studies. More recently, concerns about the uptake of science post-16 have prompted more specific studies focusing on the influences of subject choice. An examination of recent trends in A level choice show that, in England, Wales and Northern Ireland, more pupils than ever are taking A levels, but fewer, as a proportion of the cohort, are taking only science and mathematics combinations (Osborne *et al.* 1996, 1998). There is a growth in the numbers pursuing mixed A levels and also mounting evidence of a decline in the interest of young people in pursuing scientific careers (Smithers and Robinson 1988; Department for Education 1994).

To summarize, the reasons for studying attitudes to science include concerns about pupils' classroom response to science, and about subject and career choice. Before examining some of the findings of major studies and their implications, it is useful to look more closely at how attitudes are measured, and at some of the assumptions underlying different methodologies.

How are attitudes measured?

Researchers have taken a number of approaches to the measurement of attitudes to science. These include subject preference studies, where pupils have been asked to rank their liking of school subjects (Whitfield 1980; Hendley *et al.* 1995) and where attitude is inferred from relative popularity. Whitfield's analysis of 1979 International Educational Achievement data demonstrated that physics and chemistry were two of the least popular subjects post-14. A more recent study of this kind by Lightbody and Durndell (1996a, 1996b) has shown boys were far more likely to report liking science than girls. Though preference ranking is simple to use and the results easily presented and interpreted, its problem is that it is a relative scale; it is possible for a student to rank science as low, but have a more positive attitude towards it than another student who ranks it more highly. However, it can be useful if the question being asked is 'How popular is science compared with other subjects?'.

More commonly, attitudes have been measured through the use of questionnaires, which often consist of Likert scale items. Here students are asked to respond to statements such as 'science is fun', 'I would enjoy being a scientist'. Likert scales include a five-point choice consisting of 'strongly agree/ agree/not sure/disagree/strongly disagree'. Items on the scale have normally been derived from free response answers generated by students. These have then been reduced to a set of usable and reliable items, which have been piloted and further refined by statistical analysis.

Such scales have been widely used and extensively trialled and are the major feature of research in this domain. Possibly the most well known and well used is the scientific attitude inventory developed by Moore and Sutman in 1970. However, this has been criticized by Munby (1983) for the inconsistent results it produces and its lack of reliability. Moreover, a feature of this scale is that all the attitude objects are concerned with aspects of science in society and not with attitude to science as a school subject.

The emphasis on measuring attitudes through the use of questionnaires has resulted in the development of a plethora of scales which give differing degrees of emphasis to a broader range of attitude objects. More well-known examples are the instrument developed by Simpson and Troost (1982) for their large-scale study using 4500 students drawn equally from elementary, junior high and high schools in North Carolina, and the Attitudes toward Science Inventory (Gogolin and Swartz 1992) which is itself a modification of an instrument developed by Sandman (1973) to assess attitudes towards mathematics.

The problem of interpreting the significance of these multiple components of attitudes towards science has been clearly identified by Gardner (1975) who comments:

> An attitude instrument yields a score. If this score is to be meaningful, it should faithfully reflect the respondent's position on some well-defined continuum. For this to happen, the items within the scale must all be related to a single attitude object. A disparate collection of items, reflecting attitude towards a wide variety of attitude objects, does not constitute a scale, and cannot yield a meaningful score.
>
> (Gardner 1975: 12)

If there is no single construct underlying a given scale, then there is no purpose served by adding the various ratings to produce a unitary score. The best that can be done is to ensure that the components are valid and reliable measures of the constructs they purport to measure, and then look for the significance of each of these aspects. Even so, many instruments suffer from significant problems as, statistically, a good instrument needs to be internally consistent and unidimensional (Gardner 1995). Another problem identified with the use of scales and inventories on a single occasion relates to the stability of attitudes (Ramsden 1998), and the possibly erroneous assumption that attitudes are sufficiently stable to be measured at one point in time (Munby 1983). Though, once formed, attitudes may be difficult to change, there are few studies where repeated measurements of attitudes have taken place which would reliably demonstrate that the attitudes being measured are stable.

A further criticism of quantitative studies of attitudes to science has been that they provide limited understanding of the problem. Such concerns have led to studies of attitudes to science which include the use of interviews (Ebenezer and Zoller 1993; Piburn and Baker 1993; Woolnough 1994; Baker and Leary 1995). While such studies are subject to criticisms of lack of generalizability, the data do provide insights into the origins of attitudes to school science.

What do we know about attitudes to science?

Results from numerous studies of attitudes to school science (Brown 1976; Harvey and Edwards 1980; Hadden and Johnstone 1983; Smail and Kelly 1984; Simpson and Oliver 1985; Yager and Penick 1986; Johnson 1987; Doherty and Dawe 1988; Breakwell and Beardsell 1992) show that positive attitudes peak at, or before, the age of 11 and decline thereafter by quite significant amounts. Children enter secondary school with a highly favourable attitude towards science and interest in science which is eroded by their experience of school science, particularly for girls (Kahle and Lakes 1983).

Hendley *et al.*'s (1995) study of 4023 KS3 pupils in Welsh schools indicates that, out of the four core subjects, science, English, mathematics and technology, science is the least popular. This finding is confirmed by a smaller-scale qualitative study based on interviews with 190 pupils (Hendley *et al.* 1996). When asked which three subjects they liked best, science was ranked fifth out of 12 subjects. However, when asked which subjects they liked least, science emerged as the most disliked, particularly by boys. Hendley *et al.* conclude that science is a 'love–hate' subject which elicits strong feelings in pupils. Colley *et al.* (1994), in another British study, found significant gender differences among 11- to 13-year-old pupils, with girls favouring English and humanities and boys favouring PE and science.

In contrast to these results for attitudes to school science, many surveys show that students' attitudes to *science itself* are positive. Such surveys include that of the Assessment of Performance Unit (APU), and a large-scale market research survey conducted in the UK for the Institution of Electrical Engineers (The Research Business 1994). This latter study showed that students saw science as useful (68 per cent), interesting (58 per cent) and that there was no significant difference between genders. A large proportion (53 per cent) saw the relevancy of science as a reason for studying it and that it offered better employment prospects (50 per cent). Moreover, 87 per cent rated science and technology as important, or very important, in everyday life. However, this survey also showed the disparity between students' and teachers' notions of science, the former being associated with high-tech advances and social relevance, the latter with more theoretical aspects and the significant discoveries of the last century.

The contradiction between students' interest in science and their liking for school science is highlighted by the work of Ebenezer and Zoller (1993). In their study, 72 per cent of the 1564 16-year-olds interviewed indicated they thought science valuable, but nearly 40 per cent that they found science classes boring. Ebenezer and Zoller suggest that this gulf is due to the message presented by school science that science is somehow disconnected from society and that we should simply study it for its own sake.

The relationship between attitude and achievement is another key issue permeating much of the literature. Gardner's review of the research evidence offered little support for any strong relationship between these two variables. Writing somewhat later, Schibeci (1984) draws a stronger link between the two; however, he also cites studies which show no relationship. The current position is best articulated by Shrigley (1990) who argues that attitude and ability scores can be expected to correlate moderately. Weinburgh's (1995) meta-analysis of the research suggests that there is only a moderate correlation between attitude towards science and achievement, although this correlation is stronger for high and low ability girls, indicating that for these groups 'doing well' in science is closely linked with 'liking science'. Similar findings have appeared in the major study conducted by Talton and Simpson (1990).

The exception to these findings is the research of Simpson and Oliver (1990). These authors would argue that their longitudinal study shows a strong relationship between the three affective variables – attitude towards science; motivation to achieve; and the self-concept that individuals have of their own ability – and their achievement in science. In part, this may be explained by their attempt to measure 'motivation to achieve', which may be a more significant factor than attitude towards science in determining achievement. In this context, it is interesting to note the general finding that girls are always more motivated to achieve than boys. This finding might then explain why UK exams at 16 (GCSE) demonstrate that, although boys are more positively inclined towards science – more of them choose to study it and are keener to pursue the subject post-16 – their achievements are inferior relative to girls.

Factors influencing students' attitudes to science and subject choices

Research studies have identified a number of factors influencing attitudes towards science. For the purpose of this chapter, the main factors which have implications for classroom practice and subject choice are effective teaching, perceived difficulty, and gender.

Effective teaching

Several studies have pointed towards the influence of classroom environment as a significant determinant of attitude (Haladyna *et al.* 1982; Talton and Simpson 1987; Myers and Fouts 1992). Myers and Fouts found that the most positive attitudes were associated with a high level of involvement, very high level of personal support, strong positive relationships with classmates, and the use of a variety of teaching strategies and unusual learning activities. Variety as a key feature in generating interest in science also comes from the work of Piburn and Baker (1993) and the Scottish HMI report (HM Inspectors of Schools 1994) on Effective Learning and Teaching in Scottish Secondary Schools. Similar conclusions that 'school, particularly classroom, variables, are the strongest influence on attitude toward science' were drawn by Simpson and Oliver (1990) in their North Carolina study.

Evidence that effective teaching of school science is a significant determinant of attitude towards science has also been found by Woolnough whose research showed that it is a major factor in continuing with science education post-16. This finding is confirmed by the studies of Ebenezer and Zoller (1993) and Haladayana *et al.* (1982), which showed that the most important variable to affect students' attitude towards school science was the kind of science teaching they experienced. Further support for the significance of the teacher can be found in the work of Sundberg and Dini (1994) and Piburn and Baker (1993). Hendley *et al.*'s (1995) study of KS3 pupils' preferred subjects also found that the most common reasons given for liking or disliking the subject were teacher related.

In 1991 Woolnough (1994) collected data on student subject choice from a cohort of 1180 A level students. In that same year, 132 heads of science completed a separate questionnaire and 108 sixth formers and 84 staff from 12 schools were interviewed. His study identified six factors which were responsible for student choice/non-choice of the sciences. Of these, the two strongest factors were the influence of students' positive experience of extracurricular activities and the nature of in-class activities – that is the quality of their science teaching. Woolnough's work therefore supports other findings that the quality of teaching is an important determinant of attitude and subject choice.

The factors Woolnough identified as contributing to such teaching included a supply of well-qualified, enthusiastic graduate science staff (including graduates in physics and engineering), who not only have a good spread of expertise across science, but who also have individual subject loyalty. Good teaching was characterized by teachers being enthusiastic about their subject, setting it in everyday contexts and running well-ordered and stimulating science lessons. Good teachers were also sympathetic and willing to spend time, both in and out of lessons, talking with the students about science, careers and individual problems.

Perceived difficulty

Several studies (Crawley and Black 1992; Havard 1996; Hendley *et al.* 1996) have identified students' perception of science as a difficult subject as being a determinant of subject choice. In fact, Havard's investigation of the uptake of sciences at A level, albeit in only four schools, points to the perceived difficulty of science as the major factor inhibiting uptake.

Further substance to the notion that physical sciences are perceived as being difficult is provided by the analysis of the data collected by the UK Department for Education and Employment on the youth cohort for 1989, 90 and 91 using sample sizes of approximately 14,000 students for each year (Cheng *et al.* 1995). These researchers found that the most significant factors correlating with uptake of physical sciences were the grades achieved at GCSE in science and mathematics. This suggests that science is only taken by students who do well, reinforcing the notion that it is for the intelligent and is therefore perceived as difficult. Such perceptions have implications for students' self-image and career choice.

Gender

The most significant factor influencing attitudes towards science and subject choice is gender. As Gardner comments, 'sex is probably the most significant variable related towards pupils' attitude to science'. This view is supported by Schibeci's (1984) extensive review of the literature, and more recent meta-analyses of a range of research studies by Becker (1989) and Weinburgh (1995), covering the literature between 1970 and 1991. Both the latter two papers summarize numerous research studies to show that boys have a consistently more positive attitude to science than girls, although this effect is stronger in physics than in biology.

What is clear from an extensive literature on the subject, mainly as a result of a serious consideration and investigation of the problem in the 1980s, is that girls' attitudes to science are significantly less positive than boys (Harvey and Edwards 1980; Harding 1983; Kahle and Lakes 1983; Erickson and Erickson 1984; Smail and Kelly 1984; Johnson 1987; Robertson 1987; Breakwell and Beardsell 1992; Hendley *et al.* 1996). The predominant thesis offered to explain this is that it is a consequence of cultural socialization, which offers girls considerably less opportunity to tinker with technological devices and use common measuring instruments (Johnson 1987; Kahle and Lakes 1983; Smail and Kelly 1984; Thomas 1986). For instance, Kahle contends that her data show there is a gap between young girls' desire to observe common scientific phenomena and their opportunities to do so. More importantly, her data show conclusively that their science education does not remediate for this lack of experience and leads her to argue that 'lack of experiences *in* science leads to a lack of understanding *of* science and contributes to negative attitudes *to* science'. Similarly, Johnson argues from her data, measuring a range of common childhood experiences of children, that 'early established differences in the interests and activities of boys and girls result in parallel differences in their science performances'.

However, there is now some evidence beginning to appear that girls no longer hold such a stereotypical aversion to careers in science and are confident of their ability to undertake science courses (Colley *et al.* 1994; Havard 1996; Lightbody and Durndell 1996a; Whitehead 1996). In terms of achievement in science, Elwood and Comber (1995) have shown that the situation has reached a position where girls are doing as well as, if not better than, boys. These findings suggest that gender itself may now only contribute a minor part in the attribution of success. What remains an enigma is why girls choose not to pursue science even though they are competent and do believe in their capabilities to succeed.

Whitehead's (1996) research has attempted to explore, in more detail, the influence of gender stereotyping on choice. She found that, although there were significant gender distinctions within pupils' perceptions of subjects, these were not significant influences on subject choice. Girls doing mainly 'feminine' subjects, who were the focus of her study, described themselves as high on the stereotypical masculine trait of competence and were highly intrinsically motivated. Boys, in contrast, taking mainly 'masculine' subjects, were more likely to be extrinsically motivated for status, recognition and a highly paid job, describing themselves as high on the traits of competence and aggression. In general, boys are more likely to choose sex-stereotyped careers, and she suggests that this reflects a greater need to establish and strengthen their gender identity than that of girls. Hence, she suggests: 'It is not therefore that girls are under-represented in mathematics and the physical sciences, but that boys are greatly over-represented; similarly, in languages, girls appear to be over-represented in these areas only because the boys are so under-represented in them' (p. 155). Further, she comments that:

> If boys are choosing sex-appropriate subjects in order to conform to traditional notions of masculinity, then this is clearly undesirable both from the point of view of the individual, who may not necessarily be choosing those subjects at which they are most successful, and for society as a whole, as it is unlikely to gain good scientists by such a process of choice.
>
> (pp. 158–9)

Such findings would also explain why boys in boys-only schools choose more arts and humanities courses, as they are under less pressure to establish and conform to their gender identity.

Evidence which substantiates Whitehead's findings comes from work by Pauline Lightbody in Glasgow (Lightbody and Durndell 1996b; Lightbody *et al.* 1996). In a small-scale study with 106 pupils using a novel methodology to investigate career aspirations, she found no significant difference between males and females. She explains the discrepancy between this and actual career choice as a case of girls' view of science being one epitomized by the view 'We can, I can't', and that gender stereotyping is so deeply entrenched that it may not even be conscious. She argues that it is not so much that science and technology are perceived as masculine but more that the current focus of interest on technological matters is not of central interest to girls and that only a change in content and the style of teaching to show a greater interest in people will lead to a significant increase in the choice of physical sciences by girls.

This latter point is borne out by a recent, though small-scale study undertaken by Fielding (1998) into the reasons why academically capable girls are not choosing sciences and mathematics post-16. Fielding's study shows that girls who opted for science post-16 did so because they had decided upon which career to pursue. Moreover, their science-based career aspirations were all to do with people (for example medicine). Sears (1997) in his survey of A level Year 12 and 13 pupils from six comprehensive schools in England, also found that girls choosing science A levels were more likely to carry on to Higher Education in science than boys, suggesting that, having chosen science, they had a higher career commitment to science. In Fielding's study, both science and non-science A level students had similar views of the nature of scientific careers, the general consensus being that scientific jobs, other than medicine or veterinary science, would be boring and narrow. This finding suggests that girls' perceptions of careers in science are limited, which is why many do not choose science subjects post-16 even though they are competent and believe themselves capable.

Implications for classroom practice

Havard's (1996) work suggests that, in science education, a major problem lies with physics, as over 50 per cent of his sample indicated that they did not enjoy the subject at all, or very little, whereas over 60 per cent enjoyed biology a lot or quite a lot. One explanatory factor may be that physics is being taught by biology or chemistry specialists, who have little enthusiasm for the subject, which is masked by the introduction of balanced science. In such situations, teachers who lack confidence and familiarity fall back on didactic modes of teaching, and the quality of teaching and learning is impoverished (Osborne and Simon 1996).

One implication of the introduction of the National Curriculum (Woolnough 1994) has been the tendency to restrict and deskill the good science teacher, with the loss of those 'extra bits', in particular, extra-curricular activities, which contribute to good science teaching. One of Woolnough's recommendations for preserving good science teaching is that teachers should teach what they feel comfortable with, as teachers in his study were happiest and most enthusiastic when teaching their specialist subjects – a finding which clearly has implications for the nature of teaching in balanced science.

These research findings also raise the question why, despite the recurrent message of the significance of teachers and teacher styles on attitudes towards science, so little research has attempted to understand what makes for effective teaching of science in the eyes of the pupil. An interesting study undertaken in the US, providing some insight on this issue, is reported by Sheila Tobias (1990). The study aimed to explore why so many college students turn away from science in the course of their degree studies, and it involved a group of post-graduates who had successfully completed their degrees in other subjects. For a fee, the group of surrogate students were willing to re-visit introductory courses in physics and chemistry in order to audit these for the research. They each enrolled for a particular course and participated in it, attending all the lectures and doing the homework assignments and examinations. They were asked to focus their attention on what might make introductory science 'hard' or even 'alienating' for students like themselves. The seven case studies in the report reveal some common problems with those introductory courses which were alienating; the courses focused on problem-solving techniques, and lacked an intellectual overview of the subject; there were too many 'how much' questions, not enough discussion of 'how' or 'why'; pedagogy was condescending and patronizing, examinations were not challenging; there was no community or discussion and the atmosphere was competitive. One student summed up the problem as 'the absence of history or context, "the tyranny of technique", the isolation of the learner and the struggle to attend in a sea of inattentiveness' (Tobias 1990: 59).

Teachers need to be enthusiastic and knowledgeable about their subject, setting it in well-chosen contexts and running well-ordered and stimulating science lessons. The tendency for prescriptive national curricula to constrain science teaching, at the expense of interest and depth of involvement, has implications for the promotion of students' positive attitudes towards science beyond the experience of school.

The research shows that influences on subject choice post-16 are complex, and enjoyment of a subject could play a key part. The study reported by Tobias found that students did not choose science because they had a fear of cheating themselves of a 'well-rounded liberal education'. This finding is similar to that of Fielding (1998), who showed that those doing non-science subjects were often unsure as to what career they might go into, and seemed happier by the idea of keeping their options open – thinking that, if they studied science, they would have to end up as scientists. Her study also showed enjoyment as a key factor in subject choice at A level. Sears (1997) also found that students from science courses (double award in the UK), as opposed to separate sciences, considered themselves undecided about long-term career options, and he suggests that this reflects the reduction of specialism at 14. With more options open to students and less need to specialize early, it is even more crucial that enjoyment of science becomes a key factor if students are to pursue science post-16.

One would hope that the intellectual challenge and involvement experienced by the group of girls studying the electrolysis of brine could be sustained, reinforced and transferred beyond their study of GCSE chemistry.

Finally, the function of science education goes further than the provision of future scientists and service to the institution of science. The development of scientific literacy is essential to a participatory democracy. The decline in the study of science post-16, for all groups of students, indicates that we are failing to convince children that science is the most significant achievement of Western civilization. The message of this research is that the central question that teachers need to ask of their practice is 'how can we make science more appealing?'.

References

*Ajzen, I. and Fishbein, M. (1980) *Understanding Attitudes and Predicting Social Behaviour*. Englewood Cliffs, NJ: Prentice Hall.

Baker, D. and Leary, R. (1995) Letting girls speak out about science, *Journal of Research in Science Teaching*, 32(1): 3–27.

Becker, B.J. (1989) Gender and science achievement: a re-analysis of studies from two meta-analyses, *Journal of Research in Science Teaching*, 26: 141–69.

Breakwell, G.M. and Beardsell, S. (1992) Gender, parental and peer influences upon science attitudes and activities, *Public Understanding of Science*, 1(2): 183–97.

Brown, S. (1976) *Attitude Goals in Secondary School Science*. Stirling: University of Stirling.

Cheng, Y., Payne, J. and Witherspoon, S. (1995) *Science and Mathematics in Full-time Education after 16: England and Wales Youth Cohort Study*. London: Department for Education and Employment.

Colley, A., Comber, C. and Hargreaves, D. (1994) School subject preference of pupils in single sex and co-educational secondary schools, *Educational Studies*, 20: 379–86.

Crawley, F.E. and Black, C.B. (1992) Causal modelling of secondary science students' intentions to enrol in physics, *Journal of Research in Science Teaching*, 29(6): 585–99.

Crawley, F.E. and Coe, A.E. (1990) Determinants of middle school students' intentions to enrol in a high school science course: an application of the theory of reasoned action, *Journal of Research in Science Teaching*, 27(5): 461–76.

Dainton, F.S. (1968) *Council for Scientific Policy (1968) Inquiry into the Flow of Candidates in Science and Technology into Higher Education*. London: HMSO.

Department for Education (1994) *Science and Maths: A Consultation Paper on the Supply and Demand of Newly Qualified Young People*. London: Department for Education.

Doherty, J. and Dawe, J. (1988) The relationship between development maturity and attitude to school science, *Educational Studies*, 11: 93–107.

Ebenezer, J.V. and Zoller, U. (1993) Grade 10 students' perceptions of and attitudes toward science teaching and school science, *Journal of Research in Science Teaching*, 30(2): 175–86.

Elwood, J. and Comber, C. (1995) Gender differences in 'A' level examinations: the reinforcement of stereotypes. Paper presented as part of the symposium – A New ERA? New Contexts for Gender Equality: BERA conference.

Erickson, G. and Erickson, L. (1984) Females and science achievement: evidence, explanations and implications, *Science Education*, 68: 63–89.

Fielding, H. (1998) The understandable choices? Unpublished BSc dissertation, King's College London, University of London.

*Gardner, P.L. (1975) Attitudes to science, *Studies in Science Education*, 2: 1–41.

*Gardner, P.L. (1995) Measuring attitudes to science, *Research in Science Education*, 25(3): 283–9.

Gogolin, L. and Swartz, F. (1992) A quantitative and qualitative inquiry into the attitudes toward science of nonscience college majors, *Journal of Research in Science Teaching*, 29(5): 487–504.

Hadden, R.A. and Johnstone, A.H. (1983) Secondary school pupils' attitudes to science: the year of erosion, *European Journal of Science Education*, 5: 309–18.

Haladyna, T., Olsen, R. and Shaughnessy, J. (1982) Relations of student, teacher, and learning environment variables to attitudes to science, *Science Education*, 66(5): 671–87.

Harding, J. (1983) *Switched Off: The Science Education of Girls*. New York: Longman.

Harvey, T.J. and Edwards, P. (1980) Children's expectations and realisations of science, *British Journal of Educational Psychology*, 50: 74–6.

Havard, N. (1996) Student attitudes to studying A-level sciences, *Public Understanding of Science*, 5(4): 321–30.

Hendley, D., Parkinson, J., Stables, A. and Tanner, H. (1995) Gender differences in pupil attitudes to the national curriculum foundation subjects of english, mathematics, science and technology in Key Stage 3 in South Wales, *Educational Studies*, 21(1): 85–97.

Hendley, D., Stables, S. and Stables, A. (1996) Pupils' subject preferences at Key Stage 3 in South Wales, *Educational Studies*, 22(2): 177–87.

HM Inspectors of Schools (1994) *Effective Learning and Teaching in Scottish Secondary Schools: The Sciences*. Edinburgh: The Scottish Office Education Department.

Johnson, S. (1987) Gender differences in science: parallels in interest, experience and performance, *International Journal of Science Education*, 9(4): 467–81.

Kahle, J.B. and Lakes, M.K. (1983) The myth of equality in science classrooms, *Journal of Research in Science Teaching*, 20: 131–40.

Klopfer, L.E. (1971) Evaluation of Learning in Science, in B.S. Bloom, J.T. Hastings and G.F. Madaus (eds) *Handbook of Formative and Summative Evaluation of Student Learning*. London: McGraw-Hill Book Company.

Koballa Jr., T.R. (1988) The determinants of female junior high school students' intentions to enrol in elective physical science courses in high school: testing the applicability of the theory of reasoned action, *Journal of Research in Science Teaching*, 25(6): 479–92.

Lightbody, P. and Durndell, A. (1996a) Gendered career choice: is sex-stereotyping the cause or the consequence? *Educational Studies*, 22(2): 133–46.

Lightbody, P. and Durndell, A. (1996b) The masculine image of careers in science and technology – fact or fantasy? *British Journal of Educational Psychology*, 66(2): 231–46.

Lightbody, P., Siann, G., Stocks, R. and Walsh, D. (1996) Motivation and attribution at secondary school: the role of gender, *Educational Studies*, 22(1): 13–25.

Moore, R.W. and Sutman, F.X. (1970) The development, field test and validation of an inventory of scientific attitudes, *Journal of Research in Science Teaching*, 16: 217–22.

*Munby, H. (1983) Thirty studies involving 'Scientific Attitude Inventory': what confidence can we have in this instrument? *Journal of Research in Science Teaching*, 20(2): 141–62.

Myers, R.E. and Fouts, J.T. (1992) A cluster analysis of high school science classroom environments and attitude toward science, *Journal of Research in Science Teaching*, 29(9): 929–37.

Norwich, B. and Duncan, J. (1990) Attitudes, subjective norm, perceived preventive factors, intentions and learning science: testing a modified theory of reasoned action, *British Journal of Educational Psychology*, 60: 312–21.

Oliver, J.S. and Simpson, R.D. (1988) Influences of attitude toward science, achievement motivation, and science self concept on achievement in science: a longitudinal study, *Science Education*, 72(2): 143–55.

Ormerod, M.B. and Duckworth, D. (1975) *Pupils' Attitudes to Science*. Slough: NFER.

Osborne, J. and Simon, S. (1996) Primary science: past and future directions, *Studies in Science Education*, 27: 99–147.

Osborne, J.F., Driver, R. and Simon, S. (1996) Attitudes to science: a review of research and proposals for studies to inform policy relating to uptake of science. Unpublished review, King's College London.

*Osborne, J.F., Driver, R. and Simon, S. (1998) Attitudes to science: issues and concerns, *School Science Review*, 79(288): 27–34.

Piburn, M.D. and Baker, D.R. (1993) If I were the teacher . . . qualitative study of attitude toward science, *Science Education*, 77(4): 393–406.

Potter, J. and Wetherall, M. (1987) *Discourse and Social Psychology: Beyond Attitudes and Behaviour*. London: Sage Publications.

Ramsden, J.M. (1998) Mission impossible? Can anything be done about attitudes to science? *International Journal of Science Education*, 20(2): 125–37.

Robertson, I.J. (1987) Girls and boys and practical science, *International Journal of Science Education*, 9(5): 505–18.

Sandman, R.S. (1973) The development, validation, and application of a multidimensional mathematics attitude instrument. Unpublished PhD, University of Minnesota.

*Schibeci, R.A. (1984) Attitudes to science: an update, *Studies in Science Education*, 11: 26–59.

Sears, J. (1997) Children's attitudes to science and their choices post-16. Conference paper, University of York BERA Conference.

Shaw, M.E. and Wright, J.M. (1968) *Scales of Measurement of Attitude*. New York: McGraw Hill.

Shrigley, R.L. (1990) Attitude and behaviour are correlates, *Journal of Research in Science Teaching*, 27(2): 97–113.

Simpson, R.D. and Oliver, J.S. (1985) Attitude toward science and achievement motivation profiles of male and female science students in grades six through ten, *Science Education*, 69(4): 511–26.

Simpson, R.D. and Oliver, J.S. (1990) A summary of the major influences on attitude toward and achievement in science among adolescent students, *Science Education*, 74(1): 1–18.

Simpson, R.D. and Troost, K.M. (1982) Influences of commitment to and learning of science among adolescent students, *Science Education*, 69(1): 19–24.

Smail, B. and Kelly, A. (1984) Sex differences in science and technology among 11 year old schoolchildren: II – affective, *Research in Science and Technology Education*, 2: 87–106.

Smithers, A. and Robinson, P. (1988) *The Growth of Mixed A-levels*. Department of Education, University of Manchester.

Sundberg, M.D. and Dini, M.L. (1994) Decreasing course content improves student comprehension of science and attitudes toward science in freshman biology, *Journal of Research in Science Teaching*, 31(6): 679–93.

Talton, E.L. and Simpson, R.D. (1987) Relationships of attitude toward classroom environment with attitude toward and achievement in science among tenth grade biology students, *Journal of Research in Science Teaching*, 24(6): 507–25.

Talton, E.L. and Simpson, R.D. (1990) A summary of major influences on attitude toward and achievement in science among adolescent students, *Science Education*, 74(1): 1–18.

The Research Business (1994) *Views of Science Among Students, Teachers and Parents.* London: Institution of Electrical Engineers.

Thomas, G.E. (1986) Cultivating the interest of women and minorities in high school mathematics and science, *Science Education*, 73(3): 243–9.

Tobias, S (1990) *They're Not Dumb, They're Different. Stalking the Second Tier.* Tucson, AZ: Research Corporation.

*Weinburgh, M. (1995) Gender differences in student attitudes toward science: a meta-analysis of the literature from 1970 to 1991, *Journal of Research in Science Teaching*, 32(4): 387–98.

Whitehead, J.M. (1996) Sex stereotypes, gender identity and subject choice at A level, *Educational Research*, 38(2): 147–60.

Whitfield, R.C. (1980) Educational research and science teaching, *School Science Review*, 60: 411–30.

*Woolnough, B. (1994) *Effective Science Teaching.* Buckingham: Open University Press.

Yager, R.E. and Penick, J.E. (1986) Perception of four age groups toward science classes, teachers, and the value of science, *Science Education*, 70(4): 355–63.

Part II
The science department

Good practice in science teaching increasingly depends upon the corporate endeavour of the whole science department, and not just the behind-closed-doors performances of one or two high flyers. So, it is fitting that the second part of this book, on what research has to say about good practice in science teaching, looks at corporate matters.

'Management matters' would be a suitable *double entendre* for Chapter 8 authored by Justin Dillon. There he reviews evidence on what matters in achieving effective management, and the fact that it does matter in terms of making the department effective. This chapter stands alone in being about the process of management rather than about issues of the work, tasks and timeline schedules that are to be managed.

There then follow four chapters that introduce issues for which a corporate response is suitable. Julian Swain's Chapter 9 invites us to look again at summative assessment and think about how information gained from summative assessment might be used to improve programmes of study.

Philip Adey's Chapter 10 invites the reader to re-consider the evidence we have on cognitive growth. Philip's name is often associated with the Cognitive Acceleration through Science Education (CASE) project. CASE takes a strong line on how cognitive acceleration is best achieved through a whole department strategy. Schools join the CASE the project as a whole department – not as individual teachers.

Then, comes a chapter on differentiation and progression. Differentiation is now a buzz-word that permeates teacher-speak. Ofsted inspectors look for differentiation; heads of science try to write it into tasks for action plans; NQTs and student teachers are made to feel guilty about not introducing more differentiated work into their lesson plans; parents demand that work be differentiated to help their children learn more effectively. To be able to differentiate effectively, one does need some idea of what progression looks like. Yet, progression and differentiation are the obverse and reverse of the same coin. Chris Harrison, Shirley Simon and Rod Watson have jointly authored Chapter 11 on progression and differentiation. Perhaps, more than anything else, this chapter underlines the need to work together to achieve good practice in science teaching.

The development and introduction of information technology within school science is one example of a whole departmental issue *par excellence*. No individual teacher can command the resources needed to tool-up their science teaching. The science department needs to do that as a collective enterprise. But the collective enterprise needs to go beyond the hardware, beyond the software and on to the effective use of IT in science teaching on a near daily basis. This is best achieved corporately. Margaret Cox's Chapter 12 looks at what evidence there is for the effective use of IT in science education.

8 Managing the science department

Justin Dillon

This book emphasises throughout what research tells us about effective teaching and learning. However, without an understanding of effective management, putting this new knowledge into practice is likely to prove difficult – if not impossible. This chapter looks at what research tells us about effective management, particularly at the departmental level.

In 1997 the English and Welsh schools' inspectorate, Ofsted, reported on what it regarded as good practice at the departmental level, based on its observations of schools around the country. According to Her Majesty's Chief Inspector, Chris Woodhead, Ofsted found that:

- about one-fifth of schools have weaknesses in middle management which frustrate developments;
- too many heads of department (HODs) take the narrow view that their responsibility is for managing resources rather than people;
- the quality of curriculum development within a subject department is dependent upon the energy and leadership of the head of department and varies considerably; and
- monitoring to see that agreed procedures are being used and evaluated to discover their effects on the performance of pupils is poorly developed.

(HMCI 1997: 1)

The increased focus on standards, accountability and effectiveness during the 1980s and 1990s has led to greater use of techniques such as appraisal, target setting, development planning and value-added measurement by managers. However, despite the massive changes in education since the mid-1980s, one still encounters wide variation in practice in science departments.

From my own work with heads of science departments (HoDs), the evidence is that the *personal* aspects of the job are the most difficult and demanding.

One of the most challenging aspects of managing a department is trying to keep the respect of one's colleagues, while simultaneously keeping the respect of one's superiors. How, for example, do you get your colleagues, many of whom you might consider as your friends, to implement the senior management's new ideas when they think they are doing a good job already? Recognizing these and other tensions, this chapter sets out to provide aspiring, new and experienced heads of department with ideas, evidence and frameworks to help them to make sense of their own situation rather than to convince them that there is one way to be effective. The chapter focuses on some common issues where research evidence might offer some comfort, or challenge existing orthodoxies. These are:

- understanding the department – where does one begin?
- tasks, roles and responsibilities – what should HoDs be doing?
- the tension between the separate sciences – is balanced science taught by balanced scientists?
- leadership – how do you choose an effective style?
- teamwork – what makes a good team?
- decision-making – how can you avoid making a 'bad' decision?
- managing change – how can change be managed smoothly?
- teacher development – what do we know works?

Understanding the department

Why are schools organized as they are? Fayol (1916) defined the role of management as being 'to forecast and plan, to organize, to command, to co-ordinate and to control'. Although based on anecdotal experience rather than on systematic study, the simplicity of these and other early ideas led to their wide dissemination and acceptance both in civilian and military organizations. However, research into how organizations actually worked in practice challenged these simplistic ideas.

Mayo and colleagues working at the Hawthorne Plant of the Western Electrical Company in Chicago concluded that productivity is more affected by the *quality* of relationships between managers and the managed than by the effectiveness of the administrative procedures (Mayo 1945). The 'Hawthorne Effect' – innovations are effective for a short time only – is common throughout education, business and commerce. No matter how well planned an innovation might appear, neglecting people in management almost guarantees failure. Recognizing this phenomenon, Barnard (1948) introduced the idea of 'informal organizations', which he defined as the 'aggregate of the personal contacts and interactions and associated groupings of people' in a formal organization. Barnard advocated management manipulation of informal organizations by leaking information and by the use of friendship

ties – tactics that seem familiar 60 years or more on. Such 'informal organizations' work inside formal structures, which are common throughout
schools.

In the UK, many institutions, including schools and colleges, were affected
by the experience of their staff during the Second World War:

> In the period immediately after 1944, heads and administrators who
> had thrived in the hierarchical environment of the armed forces wished
> to see their new armies divided into clearly defined cohorts, with them-
> selves at the top of the pyramid and with the head of department cor-
> responding to the company commander.
>
> (Bayne-Jardine and Hannam 1972: 29)

In the 1960s and 1970s, comprehensive education led to the creation of
large schools, resulting in management issues undreamed of before the war.
However, as the saying goes, 'big is not always better' and even charismatic
heads found that there was a need to delegate responsibilities:

> Further progress towards full comprehensive education in the compre-
> hensive schools would now seem to depend very much upon the solu-
> tion of organisation and management problems in these schools. In this
> development, heads of departments have a crucial part to play, because
> it is through them that school policies become classroom practice.
>
> (Bailey 1973: 58)

During the post-war years, a range of historical, economic and cultural
factors have affected the ways in which schools are organized and managed,
and no one theory or model is adequate to explain or predict how schools
operate. Stephen Ball, writing before the 1988 Education Reform Act came
into force, summed up the tensions that heads of department would find
throughout the 1990s.

> In the contemporary jargon, heads of department are 'middle man-
> agers', with all the implications of 'line' responsibility that that suggests.
> It may be that baronial politics and the feudal relationships through
> which they currently work will be replaced by the bureaucratic proced-
> ures of management theory. On the other hand, the pristine language
> of management may only serve to obscure the real struggles over policy
> and budgets – who gets what, when and how?
>
> (Ball 1987: 237)

Although all schools are different, there are patterns and trends. Tony
Bush (1986) argued that management in relatively small, poorly resourced
departments, which are subject to innovation stress (a change in the National
Curriculum, for example), may be better understood using what he calls
'political' and 'ambiguity' models (i.e. by focusing on the personalities rather

than on the structures and systems). Conversely, large, well-resourced departments, not faced with externally imposed changes, may be better understood using 'formal' models (i.e. focusing more on the structures and systems than on individuals). Whatever the size of a department, the value of a good head of department cannot be underestimated.

Lesson 1

The skills and expertise of HoDs, their capacity to make sense of change for or with colleagues, are crucial resources, and a significant point of variation in the engagement of a department with National Curriculum texts.

(Ball and Bowe 1992: 103)

Tasks, roles and responsibilities

Middle managers are defined by the tasks that they carry out, by their official responsibilities (their accountability to pupils, parents, their staff and their managers), by the roles that they portray to others (leader, manager), and by their financial rewards. In addition, they also rely on their personal knowledge and skills in the day-to-day running of the department – managing people and problems (their roles).

How does school management change when governments engage in system-wide overhauls, such as that engineered in the UK in recent years? On the surface, heads of department are still involved in the traditional tasks of management – planning, implementing and evaluating changes: a new syllabus here, choosing a textbook there. However, at the micro-level, the job is very different now from what it was in the 1970s and 1980s, with more emphasis now on the tasks of appraisal, classroom observation and target-setting.

Wallace and Weindling (1997), surveying recent research, state that changes in working relationships in schools have resulted from the increased *responsibility* placed on middle managers, the growth in mutual *dependence* between the managed and the managers, and the widening range of *ethical dilemmas* confronting middle management.

Lesson 2

HoDs should be aware of the changes in their responsibilities, and should focus on the tasks that define their roles within the changing management in schools. To do this effectively, they need to understand the personalities, needs and skills of the people they manage.

Early writers on running a department focused primarily on the roles of the HoD. Bayne-Jardine and Hannam (1972), adapting Michael Marland's ideas (1971), identified eight areas including advocating, administering and supporting colleagues. Dunham, using informal comments from a survey of 92 HoDs, identified a number of functions most of which involve interpersonal skills but which rarely do more than hint at ethical or moral issues:

- communicating with the head;
- communicating with the other departments or with the pastoral organization;
- communicating with the teachers in the department either as individuals or in small groups;
- communicating with parents;
- administration, including planning, organizing and budgeting;
- teaching;
- staff selection.

(Dunham 1978: 47)

Beware! Lists such as these serve as normative descriptions of the type of jobs that managers *should* be doing. They do not provide information about how to do them better or about which are the most important factors. Descriptions of the 'ideal' HoD have begun to be supplemented with a broader typology of characteristics as researchers spent time looking at how real schools 'worked'.

Bayne-Jardine and Hannam claim that in 'subject departments managerial skills have rarely been considered when the tasks of heads of departments are analysed' (1972: 30). They went on to offer a number of 'pathologies' of middle managers that, they say, 'could affect the working of a department and make curriculum development and innovation difficult':

- the 'ritualist' head of department frantically hides behind a mass of detail . . .;
- the 'neurotic' worries ineffectively about the problems of carrying a theory into practice . . .;
- the 'clever young man' who has been promoted over the heads of a number of colleagues . . .

Other research identifies factors causing managers to be less effective. In a survey in the late 1970s, HoDs identified stressful situations as being caused by:

- lack of time;
- problems with working with other HoDs;
- lack of continuity within the faculty;
- role conflict;
- diversity of the role;
- frustration, e.g. in not being able to gain promotion;
- anxiety.

(Dunham 1978: 45–6)

Ten years later, Earley and Fletcher-Campbell's substantial study of HoDs and faculty heads towards the end of the 1980s, found that 'lack of time' was still a key factor (Earley and Fletcher-Campbell 1989). Whether or not a decrease in contact time (i.e. teaching) would have resulted in HoDs spending more time managing staff, the authors identified it as an issue, and their study does recognize a general reluctance among HoDs to embrace fully all aspects of the managerial role. Bailey (1973) argued that the head of department 'is paid primarily to delegate' extra work not to do it. We all recognize the pressure on managers who feel that they need to be seen to be busier than everyone else for longer periods of time.

Earley and Fletcher-Campbell's study (1989) also identified qualities that staff at all levels associated with effective department heads:

- team leadership;
- personality;
- management style;
- accessibility and proximity;
- communication;
- consultation;
- support of new teachers;
- administration;
- ethos.

Similarly, Ofsted's report on middle management contains a list of key characteristics of well-managed subject departments (HMCI 1997). Bearing in mind the caveat above, these lists *might* serve to focus managers' attention on the need to know what levers to pull and which knobs to twist: the mechanic needs to understand what makes the engine work as well as what the individual parts are called. Where can one learn to do these tasks better? The most obvious answer is from other, more experienced, HoDs. However, few HoDs spend time talking to other HoDs about managing – too often we assume that we are competent to do something simply because we are paid to do it. Learning to be a manager is a lonely occupation in most schools.

Lesson 3

HoDs tend not to do all their key tasks. Instead, they tend to focus too much on administration. In situations where this is the case, managers need to delegate tasks and spend more time finding out what is happening in the department, and analysing how to make teaching and learning more effective in the classroom. Reading the rest of this book is a start.

Leadership styles

> Middle management at its best fits leadership styles grounded in an organic organisational philosophy where middle managers function as *communicating links* between the senior management team and the teachers, as *support agents* for teachers in their daily work and as *stimulating forces* to the improvement of teaching and learning in the school.
>
> (Brown *et al.* 1997: 6, emphasis added)

Management involves getting things done. Leadership involves getting things done in a particular way. Leadership and management are interrelated but there are many ways to be a good leader. We all have experience of working with people who would do things differently from us and are just as successful. Instead of looking for a single set of characteristics of good leaders, it is more useful to look at what leadership actually involves – because, as Immergart (1988: 262) found, 'effective leaders exhibit a repertoire of styles and ... style is related to situation both context and task'. As a result, 'the HoD may well find himself/herself leading the departmental team in one way and working with individuals in quite different ways' (Turner and Bolam 1997: 3).

Most of the research into leadership has focused on the leader rather than the led. However, Hersey *et al.* (1996: 190) point out the important part played by followers who are 'vital not only because individually they accept or reject the leader but because as a group, they actually determine whatever power the leader may have'. Bennett (1995: 17) found that "strong leadership and weak fellowship" may lead to uncritical acceptance of the interpretation put forward by the leader; "strong followership", in which the ideas of the leader are scrutinised critically and debated, may produce some negotiation, but the eventual interpretation of what is needed will probably be stronger'.

Lesson 4

The context of leadership is important as well as the individual characteristics of a particular HoD. Appointing a successful HoD from one school is not a guarantee that they will be successful with another team. Beware of 'strong' leaders working with 'weak' followers.

The influence of science on science departments

Science teachers are not a homogeneous group. Science departments usually contain a mixture of biologists, chemists and physicists. Musgrove regarded subjects (in general, not just in school) as 'communities of people competing and collaborating with one another, defining and defending their boundaries, demanding allegiance from their members and conferring a sense of identity upon them ...' (1968: 10). Bayne-Jardine and Hannam offer cautionary words of advice:

the head of department has a natural vested interest in his subject. If the place of his subject in the curriculum is threatened then he will feel that the basis of his power is questioned ... Such developments as Integrated Science ... are increasing. If the head of department takes part in such developments he will find that he not only has a management problem within the school organisation but also he may be undermining the basis of his professional position.

(Bayne-Jardine and Hannam 1972: 26 and 30)

During the 1980s and 1990s, a series of studies indicated that teachers of different subjects exhibited characteristics which might be described as 'tribal' or 'sub-cultural' behaviour in terms of grouping pupils (Ball 1981), and in terms of the different views of the goals teachers have for their pupils (Grossman and Stodolsky 1993). Grossman and Stodolsky also point to the role played by teachers' subject background in 'filtering and shaping the way teachers plan their work and interact with students' (quoted in Wildy and Wallace 1995: 1).

Science departments are often faced with a number of problems which are external to the school but which need to be understood by HoDs. First, school science is often perceived as a whole by outsiders who are ignorant of the differing cultures, histories and traditions of the separate sciences and of the strains placed on science departments by the move to integrated science, particularly during the 1980s in the UK. Secondly, teachers with a limited knowledge of science often end up teaching science, particularly to younger students in secondary schools. Thirdly, the gendered nature of science – physics has been seen as a 'masculine' subject – biology as 'feminine' – has often resulted in a gender imbalance in subject choices and in the gender mix of departments.

Lesson 5

Middle managers would do well to consider their colleagues' view of their subject and encourage teachers to examine their own positions on the role of management and leadership. However, people are often unaware of the nature of their subject history and sometimes hold erroneous views of the history and the utility of their subject.

Groups and teams

Teachers are often reluctant to see committees or working parties as effective ways of doing things if too much time is taken up receiving information and not enough time spent on getting things done. Research into teams and teamwork provides a rich, though confusing, set of ideas. Much of the work on teams emanates from the business world and, depending on one's attitude towards commerce and industry, care needs to be taken to learn the lessons that should be learned rather than to ignore or accept ideas uncritically.

In the early 1980s, the work of Meredith Belbin was particularly influential. Belbin studied teams and identified roles that he argued were important for efficient teamwork. Belbin's list of team roles, which has gone through several revisions, is frequently cited in books and courses on group management. Some people might enjoy recognizing themselves as a 'Resource Investigator' – someone who is 'creative, imaginative, unorthodox'; someone who 'solves difficult problems'. They might also recognize their 'allowable weakness' – 'Ignores details. Too preoccupied to communicate effectively'. However, others might object to being described as 'Lacks drive and ability to inspire others. Overly critical' (Belbin 1998). Belbin argued that individuals can act out more than one of the roles in particular circumstances – something often ignored when his work is quoted.

Lesson 6

Groups might benefit from letting people work in different ways – 'bringing out the best' in individuals benefits the whole team. Good HoDs can identify gaps in their team and find ways to fill them – either by changing role themselves or by encouraging others to do so.

(Gold 1998: 22)

Michael Kirton argues that teams that contain a spread of people on a spectrum from 'adapters' (who fit innovations into their ways of working) to 'innovators' (who throw out their normal ways of working when something new comes along) are likely to perform better than teams where everyone is at one end of the spectrum, or in teams that are too polarized (Kirton 1989). Many of these psychological constructs, such as innovation–adaptation are measurable using an inventory – usually a questionnaire – which has been constructed to identify roles or personality traits. Although undergoing a renaissance in some parts of the business world, teachers have generally been sceptical of such techniques.

Lesson 7

Seeing teams as evolving groups of people with collective wisdom and capability, and yet dependent on individual characteristics, helps to explain why it takes time for groups to begin to work well. However, attempts to pigeon-hole staff into the roles, and assume that they are fixed, are usually the sign of poor management.

Decisions

It is the decisions that are taken, or not taken, that play a key part in determining departmental effectiveness. Ball and Bowe point to the changing nature of decisions that departments now have to make:

The new process of curriculum development emerging out of the National Curriculum involves the replacement of local decisions based on direct experience with general structures based on assumptions about 'normal' pacing and 'levels' of difficulty.

(Ball and Bowe 1992: 103)

The department meeting is regarded by some as fundamental to the process of decision-making. Michael Marland argues, rightly in my opinion, for a decrease in administration during meetings: 'This leaves the formal regular meeting . . . free from administration . . . and available for its central purpose: educational discussion' (Marland 1971: 28). In many schools, departments take radical decisions while others might, in effect, decide not to resolve anything at the moment. Penelope Weston describes what happened in one science department: 'there was an element of laissez-faire about [the] episode of decision-making, so that the decision *against* integrated science in the end almost took itself' (Weston 1979: 255). Both approaches can be effective mechanisms so anyone seeking to improve the quality of decisions, and decision-making, needs to be wary of ideal visions of the process.

Some years ago, the Association for Science Education (ASE) commissioned research into decision-making in school science departments (Hull and Adams 1981). The researchers argued that 'the greater the degree of participation in [major decisions], notwithstanding limiting factors such as lack of experience within the department or shortage of time to devote to discussions, the more successful the whole process is likely to be' (Hull and Adams 1981: 42). However, I can find little evidence to support this point of view. It has been found (Brown 1990) that *usually* groups perform less well on intellectual tasks, such as decision-making, than individuals. So, why do schools bother to have departmental meetings? March comments that:

Theories of choice assume that the primary reason for decision making is to make choices. They ignore the extent to which decision making is a ritual activity closely linked to Western ideologies of rationality. In actual decision situations, symbolic and ritual aspects are often a major factor.

(March 1982: 37)

It is certainly not the case that group decisions are the average of individual pre-discussion positions as was thought until about 1960 (Brown 1990). Several studies have found that groups tend to shift *towards* an extreme position beyond the average pre-discussion position.

Lesson 8

The act of meeting performs other functions than simply discussing and deciding. Cancelling meetings can have an impact on staff perceptions of the value of their 'participation' in the information flow or in the decision-making process. Colleagues' non-attendance at departmental meetings may well be indicators of conflicting loyalties, or indicators of a wish to avoid conflict, or to opt out of commitment to decisions.

Can we learn anything from looking at how poor decisions are made? Janis (1972) identifies five factors that led to 'bad' decisions (i.e. things turned out badly for the decision-makers) as a result of the decision:

- the group making the decision was very cohesive;
- it was insulated from information from outside the group;
- the group rarely searched systematically through alternative policy options to appraise their relative merits;
- the group was under stress to make a decision quickly;
- the group was nearly always dominated by a very directive leader.

According to Janis, these conditions led to strong pressures to conform, leading to what Janis called 'groupthink'. That is to say, the group members effectively exerted pressures on possibly dissenting individuals.

Changes in individual attitudes, especially of younger, recently qualified teachers, can also affect the decisions that departments make over a period of time. Richardson (1981) found that, at the end of their probationary year, teachers were less favourable to teacher autonomy in decision-making, and showed greater acceptance of the headteacher's decision-making authority than they had been as student teachers. This challenges the widely held view that newly qualified teachers tend to challenge existing orthodoxies.

Lesson 9

Individuals may respond to the process of decision-making in different ways. Democratic decision-making will not necessarily lead to good decisions. However, the ritual role of decision-making is important. The job of the HoD is to decide, in the light of the personalities present in a department, how much time to spend on discussing and how much autonomy to take in decision-making.

Finally, it is important to be aware of the gap between decision and implementation: 'Decisions made at departmental meetings have been forgotten or not implemented by certain staff' (Prickett 1982: 210). As Carrie Paechter wrote: 'decisions seem almost to replace actions in people's heads. It is as if by taking decisions individuals come to believe that they have carried out the action it implies' (Paechter 1995: 56).

Managing change

So what should managers do that will raise standards of teaching and learning that will be sustained? Many of the strategies suggested in the other chapters in this book will be useful to the majority of teachers. What is required by HoDs is an effective approach to implementing them. McLaughlin, who analysed a major US study into change, reports that effective strategies should:

- be concrete, teacher-specific and involve extended training;
- involve classroom assistance from local staff;
- use teacher observation of similar projects in other classrooms, schools or districts;
- utilize regular project meetings that focused on practical issues;
- ensure teacher participation in project decisions;
- support local development of project materials;
- encourage principals' participation in training.

(McLaughlin 1990: 12)

Fullan (1991), and others, argue that implementation involves three steps: initiation, implementation and institutionalization. In order to embed change in an organization, it is necessary to understand how the organization works, and this can only be done by talking to people outside the department as well as inside. Having collected enough information about how the department is perceived by insiders and outsiders, the time comes to take action.

Lesson 10

HoDs need clear ideas about what is needed in their department, and they need to have concrete strategies to implement. Implementation needs to be supported by the senior management and should involve teachers working together over long periods.

Teacher development

As it is teachers who can make the biggest difference to student learning, the emphasis of an HoD's efforts needs to be on teacher development. Without addressing what teachers do, say, think and use, efforts to raise standards will fail. However, teacher development is complex and, as yet, incompletely understood. Progress has been made by learning from successful projects, and some models of teacher development appear to be useful. Often the message is that teachers need to find the time and external motivation in order to reflect on their progress.

In England and Wales, the Teacher Training Agency (TTA) has outlined standards for subject leaders (that is to say, heads of department and primary teachers responsible for a subject) 'which set out the knowledge, under-standing, skills and attributes' which 'define expertise in subject leadership and are designed to guide the professional development of teachers aiming to increase their effectiveness as subject leaders' (TTA 1997: 2). 'Key out-comes of subject leadership' include 'teachers who: work well together as a team; support the aims of the subject . . . [and] are dedicated to improving standards of teaching and learning' (p. 4). These 'standards' are typical of the model of teacher development held by policy-makers and are usually too vague to be helpful in the cold twilight of an in-service session.

Bell and Gilbert's (1996) model of teacher development, based on work with science teachers in New Zealand, describes three strands of develop-ment: social, professional, and personal. Social development 'involves the renegotiation and reconstructing of what it means to be a teacher'; personal development 'involves . . . constructing, evaluating and accepting or rejecting . . . the new socially constructed knowledge about what it means to be a teacher, . . . and managing the feelings associated with changing their activ-ities and beliefs about science education'; and professional development 'involves not only the use of new teaching activities in the classroom but also the development of the beliefs and conceptions underlying the actions' (Bell and Gilbert 1996: 161). This type of model, which separates the processes of teacher development from the goals, seems to me to get closer to what really goes on when groups of teachers 'develop' themselves and each other.

Mandatory appraisal for teachers was one strategy brought in in recent years in order to make teachers observe each other and talk specifically about development. However, appraisal has failed to gain universal approval in schools because it often fails to address teachers' needs adequately. The emphasis has too often been on the institution and not on the individual. A second issue has been the inappropriate and inadequate nature of profes-sional development available to teachers.

Joyce and Showers (1988) reviewed research into teacher development and summarized the different approaches used by in-service providers. These are:

- presentation of theory and description of skill or strategy;
- modelling or demonstration of the skills or models of teaching;
- practice in simulated and classroom settings;
- structured and open-ended feedback (provision of information about per-formance);
- coaching for direct application (hands-on, in-classroom assistance with the transfer of skills and strategies to the classroom).

The authors found that studies involving coaching were more likely to be successful than any other form of development. The implication for man-agers is that there must be some aspect of classroom observation, feedback and teamwork as well as simply agreeing to try something out in the privacy of one's own laboratory.

Lesson 11

Teacher development will only happen if teachers' needs are addressed directly and if teachers are able to see their own situation and choose strategies for self-improvement in a supportive environment, which involves feedback which is both positive and helpful.

Finally

The 1990s have been the decade in which raising standards and school effectiveness have become the major concern of politicians across the spectrum. However, it is important to recognize just what the research into effective schools has shown rather than to accept or reject the ideas of 'effective schools' wholesale. Sammons *et al.* (1997) undertook a multi-level statistical analysis of GCSE (16+ examinations) data from 18,000 students over 3 years in 94 schools followed by case studies of 30 departments in 6 schools and a questionnaire survey of 90 schools' heads and heads of departments of English and mathematics. The researchers concluded that the majority of departments were, in general, similar in effectiveness and that, in any one year, only around 25–30 per cent were 'more or less effective' in overall performance and subject results. This should be a salutary finding to those who believe that they have the answers to raise performance in schools.

Research offers HoDs some ideas about the complexity of their job and some insight into what might work and what might not. Someone described management as 'like herding cats' – if only it were that simple.

References

Bailey, P. (1973) The functions of heads of departments in comprehensive schools, *Journal of Educational Administration and History*, 5(1): 52–8.

Ball, S.J. (1981) *Beachside Comprehensive*. Cambridge: Cambridge University Press.

Ball, S.J. (1987) *The Micro-politics of the School*. London: Routledge.

Ball, S.J. and Bowe, R. (1992) Subject departments and the 'implementation' of National Curriculum policy: an overview of the issues, *Journal of Curriculum Studies*, 24(2): 97–115.

Barnard, C.I. (1948) *The Functions of the Executive*. Cambridge, MA: Harvard University Press.

Bayne-Jardine, C.C. and Hannam, C. (1972) Heads of departments, *Forum*, 15(1): 26, 29–30.

Belbin, R.M. (1998) *The Coming Shape of Organization*. Oxford: Butterworth-Heinemann.

*Bell, B. and Gilbert, J. (1996) *Teacher Development: A Model from Science Education*. London: Falmer.

Bennett, N. (1995) *Managing Professional Teachers*. London: Paul Chapman Publishing.

Brown, M., Boyle, B. and Boyle, T. (1997) The effect of decentralisation on the shared management role of the head of department in UK secondary schools. Conference paper, York, Annual Meeting of the British Educational Research Association (BERA).

Brown, R. (1990) *Group Processes*. Oxford: Basil Blackwell.

Bush, T. (1986) *Theories of Educational Management*. London: Paul Chapman Publishing.

Dunham, J. (1978) Change and stress in the head of department's role, *Educational Research*, 21(1): 44–7.

*Earley, P. and Fletcher-Campbell, F. (1989) *The Time to Manage?* London: NFER/Routledge.

Fayol, H. (1916) *Administration Industrielle et Générale*. Translated by C. Storrs (1949) as *General and Industrial Management*. London: Pitman.

*Fullan, M. (1991) *The New Meaning of Educational Change*. London: Cassell.

*Gold, A. (1998) *Head of Department: Principles in Practice*. London: Cassell.

Grossman, P. and Stodolsky, S. (1993) Adapting to diverse learners: teacher beliefs in context. Conference paper, Atlanta, GA, Annual meeting of the American Educational Research Association.

*Her Majesty's Chief Inspector of Schools (HMCI) (1997) *Subject Management in Secondary Schools: Aspects of Good Practice*. London: Ofsted.

Hersey, P., Blanchard, K. and Johnson, D. (1996) *Management of Organisational Behaviour: Utilising Human Resources*, 7th edn. New York: Prentice Hall.

Hull, R. and Adams, H. (1981) *Decisions in the Science Department: Organization and Curriculum*. Hatfield: Association for Science Education.

Immergart, G.L. (1988) Leadership and leader behaviour, in N.J. Boyan (ed.) *Handbook of Research on Educational Administration*. New York: Longman.

Janis, I.L. (1972) *Victims of Groupthink*. Boston, MA: Houghton Mifflin.

Joyce, B. and Showers, B. (1988) *Student Achievement through Staff Development*. New York: Longman.

Kirton, M. (1989) *Adaptors and Innovators: Styles of Creativity and Problem Solving*. London: Routledge.

McLaughlin, M. (1990) The Rand change agent study revisited: macro perspectives micro realities, *Educational Researcher*, 19(9): 11–16.

March, J.G. (1982) Theories of choice and making decisions, *Society*, November/December: 29–39.

Marland, M. (1971) *Head of Department: Leading a Department in a Comprehensive School*. London: Heinemann.

Mayo, E. (1945) *The Social Problems of an Industrial Civilisation*. Cambridge, MA: Harvard University Press.

Paechter, C.F. (1995) *Crossing Subject Boundaries: The Micropolitics of Curriculum Innovation*. London: HMSO.

Prickett, G.J. (1982) Departmental self-evaluation in practice, *School Science Review*, 64(227): 207–12.

Richardson, G.A. (1981) Student-teacher attitudes towards decision-making in schools before and after taking up first appointments, *Educational Studies*, 7(1): 7–15.

*Sammons, P., Thomas, S. and Mortimore, P. (1997) *Forging Links: Effective Schools and Effective Departments*. London: Paul Chapman Press.

Teacher Training Agency (TTA) (1997) *National Standards for Subject Leaders – Annex*. London: TTA.

Turner, C. and Bolam, R. (1997) Analysing the role of the subject head of department in secondary schools in England and Wales: towards a theoretical framework. Conference paper, York, Annual Meeting of the British Educational Research Association (BERA).

Wallace, M. and Weindling, D. (1997) *Managing Schools in the Post-Reform Era: Messages of Recent Research*. Cardiff: Cardiff University of Wales for the Economic and Social Research Council.

Weston, P.B. (1979) *Negotiating the Curriculum: A Study in Secondary Schooling*. Windsor: NFER.

Wildy, H. and Wallace, J. (1995) Science as content; science as context: working in the science department. Conference paper, San Francisco, CA, Annual Meeting of the National Association for Research in Science Teaching (NARST).

9 Summative assessment

Julian Swain

But in this world nothing can be said to be certain, except death, taxes *and assessment.*

With apologies to Benjamin Franklin

The scope of summative assessment

Assessing students is one of the oldest practices in education, and the term 'summative assessment' is often regarded as the end of any assessment for students – having a supposed finality. Classroom teachers often show little enthusiasm for summative assessment, tending to accept rather than understand its mechanism or its consequences. Although reports of experimental research on summative assessment are not common, there does exist a variety of literature where summative data have been analysed and interpreted. Both teachers and administrators should realize that such literature exists, and that it can be useful to them in shaping their actions and decisions, both for the benefit of present and future students. In this chapter, using evidence from summative assessments and research, an argument will be developed showing how summative assessment is an expanding area, worthy of more attention by teachers, more research and further development.

Students are complex and multi-faceted. Like jewels, they are sometimes best inspected facet by facet rather than viewed as a whole: assessment can attempt to profile students rather than give a single mark or grade. The data on any student, like any jewel, can be put to different uses, for example, as part of a collection which, in the case of summative assessment data, may be classrooms, schools or nations. Each student represents part of a nation's investment in the future and, both collectively and individually, they give an indication of the quality and intellectual wealth. It is this variety of perspectives, from the individual to the whole, and its multiplicity of uses that can, and should, be looked at in order to provide new insights into the performance of educational systems. Summative assessment is the means of performing this task.

The generation of summative data

Many of us tend to think of summative assessment as the end of topic test, national tests, or the terminal examination in Year 11, where the information issued is a statement of achievement for the student at that time. However, the production of summative data has become a world-wide industry. A hundred years ago tests were limited to the classroom. Now they have become globalized, and national and international tests are conducted in almost all countries of the world, comparing schools with schools, and countries with countries.

This chapter looks at the range of recent types of summative assessments and associated research by examining and discussing data from:

• local education authorities (LEAs) and schools;
• national tests;
• UK national examinations at ages 16 (GCSE) and 18 (GCE A level);
• the national programme of the Assessment of Performance Unit (APU) in the UK;
• the Third International Mathematics and Science Study (IEA (TIMSS) 1996/7, 1997).

In all of these tests, the approach is essentially bottom-up, where the individual student in the classroom provides the data for local, national and international comparisons. This might give the impression that the individual student is less important than the system of testing in which they find themselves. While this is inescapable, the research on summative data, and its applications summarized here, has its greatest value when its implications for future cohorts and individuals are examined! Each of the above areas is very broad and is capable of a fuller analysis than space permits here. Therefore, references to the fuller works are given at the end.

Performance and summative assessment

Surprisingly, the term 'summative assessment' is comparatively new. Bloom *et al.* (1971) defined summative evaluation tests as those assessments given at the end of units, mid-term and at the end of course, which are designed to judge the extent of students' learning of the material in a course for the purpose of grading, certification, evaluation of progress or even for researching the effectiveness of a curriculum. The definition that tends to be used by many teachers and educators is that it is information derived from external agencies, at a given point in time, which defines aspects of a student's or school's performance. The information derived is seen as being of little or no use to the student once it has been generated. It may have consequences for the student's future in that they may, or may not, study subjects at A level, or go to university to do a degree course. It is in complete contrast to formative assessment (see Chapter 2) which attempts to provide feedback to the student so that better learning can take place in the future, and also, to the teacher for better teaching and greater understanding of the pupils. Problems occur when the functions of formative and summative assessments are not clearly delineated and this can produce tensions between the two (Wiliam and Black 1996). This has been particularly apparent in the UK National Curriculum assessment programme, where data on schools or educational authorities are seen as being more important than the performance of individuals within these institutions or organizations.

The contribution of the individual student to the data cannot be overemphasized. In a school, each student receives an educational experience, which will be translated into particular grades or levels of achievement, when summative assessments are taken. Historically, grading pupils has undergone a number of changes. Originally, the mark out of ten, or a percentage, in some way characterized teachers' knowledge of testing. However, national examinations, over many decades, have emphasized the application of 'norms' and 'criteria' to student marks. Applying 'norm-referencing' to marks assumes that there is some underlying pattern (the normal bell-shaped curve) in the way marks or grades are distributed in a population, and that this pattern is relatively stable from year to year. It also assumes that changing the educational experience will not change the performance of the class, school, education authority and nation. Clearly, this is not the case as patterns of mark distributions do change, and yet the causes for them may not be apparent without detailed analysis. On the other hand, criterion referencing (Popham 1978), which is associated with mastery, looks at the performance of the individual student and at what they can do in defined domains, such as biology, practical work, and so on. The collection of jewels, referred to previously, may contain some gems which satisfy specific criteria such as high-quality cut, brilliance and colour, but others may have good cut but poor colour, and so on. Specifying an extended list of criteria in all educational domains, and then trying to assess them, would be difficult before extensive research was carried out. Also, the interpretation of the criteria by teachers may prove difficult (Lang 1982). Kempa and Odiaga (1984) suggested that it was difficult to compare grades derived from norm-referenced examinations with those obtained from criterion-referenced performances, on different abilities and skills. Statistical aspects of criterion-referenced assessment were subject to intense debate during the 1970s (Berk 1980), and little progress has been made in its adoption in national testing.

Some of the technical issues associated with all types of assessment have been given by Wiliam (1993). Terms such as validity – do the tests measure what they are supposed to measure? dependability – how much reliance can we place on the results? and reliability – how accurate are the results? are often misunderstood. For example, a test can be reliable in the sense that the results obtained are repeatable, or the questions correlate with total scores, but it may not be valid because it does not measure what it was intended to measure.

Local views from the school, classroom and local education authority

Schools and the classroom

Each year in the UK, and elsewhere, headteachers, school governors and local education authorities await the publication of lists of school performance defined by examinations such as the GCSE, GCE A level and tests at age 11 and 14 (key stage tests). Headteachers try to account for shortfalls or improvements in their performance during the year, so that the credibility and parental choice for the school is maintained. Yet, the idea that such tests may be susceptible to a range of errors and criticisms is never more than fleetingly considered. The general public's erroneous notion that a mark is a mark, and a grade is a grade, is something which has stood the test of time.

Typically, diagrammatic representations of science results for a school are common (Figure 9.1).

For science departments, it is important that such data are used to ask questions such as – how do the results in the department compare with others locally? What is happening over time? What can we learn from them? What do the results hide? In that patterns over time are a more effective indicator of a department's performance than those of a single year, Borrows (1997) has provided some potentially helpful ways of analysing examination results. The average level attained for Year 9 (age 14) pupils in the UK national tests (KS3) lies between 5 or 6 (on an 8-level scale). Hence, an increase in the proportion achieving these levels over time would be indicative of rising standards within the school, provided that they are not at the expense of the higher levels. The graphical analysis (Figure 9.2) shows how one school distribution, which was slightly below the national figures, changed over time, particularly at level 6, to the one where it is now.

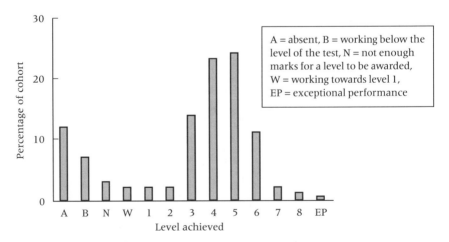

Figure 9.1   Graph showing typical display of test results

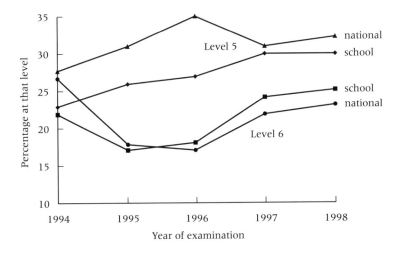

Figure 9.2 National test results for Headwood School

However, even this type of analysis can be deceptive and can hide important data. For example, if the results from the pupils' test papers are re-worked into subject domains, we might find that the biology teaching is far from adequate; or that the results of one class are much worse than another, even if they have a similar ability range. If schools are to obtain maximum information from such test results, then time must be allocated for such analysis; a cursory examination is insufficient to reveal the possible richness within the data. Statistical programmes for school use increasingly are being developed, and advisers are now beginning to help departments understand the data more fully. Only this type of detailed analysis will allow summative information to be used in a way which will help to evaluate schemes of work, the quality of teaching, and pupil understanding in the different scientific domains. Acting on such information can then help to raise the performance of future pupils.

Change can be a consequence of legislation, but internal change can be equally effective. Taking action such as setting targets within schools and involving the teachers with this target-setting process (DfEE/Ofsted 1996), or making effective use of national test data (SCAA 1997) for curriculum design and monitoring pupil progress, can all help to raise educational standards. Consequently, understanding the techniques for analysing summative assessment data and interpreting the results are becoming increasingly important skills.

Local education authorities and national data

In the UK, the role of the local education authorities (LEAs) in assessment has changed considerably during the last ten years due, in part, to four aspects (Conner and James 1996). These are:

- the changes to the assessment orders which determine what is to be assessed, by whom, and how it is to be reported;
- the influence of financial controls such as grants;
- the introduction of the national tests and tasks so that any guidance on assessment that the LEA gives to schools is usually directed to this area;
- the influence of the Office for Standards in Education (Ofsted) which looks at assessment and reporting as part of its framework for inspection.

LEAs are therefore anxious to monitor performance of their schools and compare them with national results. Typically results for LEAs are compared with national figures (Figure 9.3). These are usually distributed to schools and follow up in-service programmes are provided to try to enhance their future performance.

Murphy (1997) expresses a number of legitimate concerns over the publication of league tables: comparing results between different years rests on an assumption that the demographic characteristics are similar, which is often unjustified; comparing achievement between different subjects rests on an assumption that each subject tests similar aptitudes and abilities; and comparing schools in league tables tests an assumption that all schools start with pupils of similar ability. *All* of these assumptions are highly questionable. For instance, research by Strand (1998) with primary school test data revealed significant differences between schools' raw results, as given in the performance tables, and those schools' results which included additional measures of effectiveness of the ability of their pupils. Similarly, research has shown that performance in national public examinations is underpinned by variations in socio-economic background of pupils. Gibson and Asthana (1998) have explained how such statistics, which do not acknowledge the context of the performance, are invalid, and that policies for school

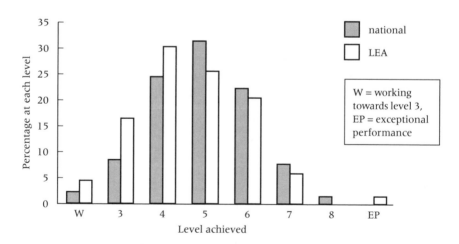

Figure 9.3 Comparision of LEA with national results

improvement must acknowledge underlying constraints. Debates over the use of published league tables will no doubt continue and research (DfEE 1995; Jesson 1997) is now beginning to be conducted in value-added measures, which will attempt to give fairer pictures of the performance of schools.

Views from the national tests

Traditionally, examinations have been developed to monitor individuals, and it is only more recently that assessment has been used to monitor national or local systems. The foundations of the national assessment system introduced by the 1989 National Curriculum stemmed from a report of a task group on assessment and testing (DES 1988). Development on new forms of national assessment at KS3 in science, mathematics, English and technology was started in 1989. 1990 was the first year in which trials were conducted, followed by further trials in 1991 (Swain 1991a, 1991b, 1992). Right-wing political pressure (Black 1994, 1998a, 1998b) then demanded that the style of these assessments should be changed to a 'pencil and paper' format rather than be administered in the classroom with practical elements by teachers. Thus 'new'-style tests were first used in 1992 and have remained in a similar format ever since in the UK.

The standards of performance are closely monitored each year and the average level for pupils at age 14 (the end of KS3) is set at between level 5 and 6. In the UK, results show that the proportion of pupils attaining level 5, and above, rose slightly in 1997 to 60 per cent, as did the proportion achieving level 6, and above, to 29 per cent. However, the overall patterns are very similar and each year shows only slight variation (Table 9.1). While the setting of the questions on these tests is criterion-referenced to the National Curriculum, the marking of the questions in the tests is not. The marking used is numerical and statistical, and normative methods are used to ensure consistency of results.

One important source of information for formative use is the annual reports produced by the testing agencies. They not only contain valuable data on the overall performance on questions and topics within papers but they also give clear messages on the implications of the findings for teaching and learning. For example, the Qualifications and Curriculum Authority (QCA 1998) report makes several references to the weaker performance of pupils, when using extended prose in answers and recommends that further opportunities be given to this in the classroom. Likewise, there are comments on specific content areas: 'Candidates continue to need more opportunities to develop their knowledge and understanding of geological changes', or 'Pupils need to be given more opportunities to consolidate and develop their knowledge on the solar system.' Teachers who choose to ignore these reports, and there is some evidence to suggest that they do (Swain 1996), will be doing a disservice to future cohorts.

Table 9.1   The national results for Key Stage 3, percentage of children achieving a particular level from 1994 to 1997

Level of performance	< 3	3	4	5	6	7	8
1994	3	9	19	28	27	9	0
1995	2	10	24	31	18	7	0
1996	2	9	26	35	17	4	0
1997	2	8	24	31	22	7	0

Source: QCA (1998).

At age 14, the UK national tests are intended to assess over 200 hours of science teaching and learning by the student, in a three-hour examination period. One of the statistics used to measure the reliability of these tests is known as Cronbach's Alpha. It is a measure of internal consistency and looks at the extent to which questions within the examination all measure the same thing. Using this technique, the national tests have been shown to have high reliability. However, while a test may be reliable, it may measure such a limited set of knowledge and competencies that its validity remains questionable.

Thus, such national tests provide only a limited sketch of pupil performance at a particular point in time. It is the teacher who is in a much better position to provide a more coherent picture. Teachers see their students at work every week and they know their strengths and weaknesses in a way that the national tests cannot measure. In the UK teachers are required to produce their own assessment of pupils' levels of achievement, which are collected separately. However, they are not required to do this until the national test results are known. Consequently, the correlation between the national distributions of levels from the tests and the teacher assessments is not unsurprisingly high (QCA 1998)! As the results of the teacher assessments are not combined with the results of national tests, there is a hidden implication that the teachers' judgements are unreliable. However, studies in Queensland in Australia (Butler 1995), suggest that teachers can provide both valid and reliable assessments. This has been achieved by the teachers working with the State developers over a period of time, and so they now have a sense of participation and ownership of the assessment.

Views of national examinations at ages 16+ (GCSE) and 18+ (GCE A level)

National views from public examinations deserve a separate section because they have been the subject of considerable discussion and analysis. Also, in the UK, there are special problems of comparability, as the administration of these examinations is not conducted by a single organization, unlike the national tests which have a single body for each key stage. Some of the more important aspects of the analysis of these summative assessments are discussed.

The numbers game and entries to 16+ examinations: a UK perspective

During the last 12 years in the UK there have been both curriculum changes and assessment changes at 16+, and the summative data show this. For example, the number of candidates taking GCSE double award science since 1989 has increased ten-fold. This increase in science entry has not been at the expense of other subjects such as English or Mathematics but at the expense of the separate science subjects themselves. Thus, comparing summative data over a number of years is totally unreliable, as similar cohorts are not being compared.

Recent target-setting and the publication of league tables has made the number of entries and grades obtained particularly important. For instance, where it is possible to have three separate awards in the sciences, instead of the single subject double award, there is a potential to maximize the number of top (A to C) passes per student, and hence raise possible positions in league tables. In such a context, education is now dominated by a market ideology (Ball 1990; Apple 1992), whose function is the production of a labour force which will sustain the economic growth of the nation and whose performance must be monitored through the use of inspections and examination results. These, in turn, lead to competition between schools. As a result, schools can seek to implement spurious curriculum changes that are of little benefit to pupils, such as allowing three separate sciences, but which will enhance their market position within the community.

Subject difficulty

A report by Fitz-Gibbon and Vincent (1994) found that differences in subject difficulties at age 18 (A level in the UK) ranged from about a third of a grade up to a grade and a quarter. In this type of study, the grades achieved by a candidate in one subject are compared with the grades achieved in another subject. These subject pairs are then compared. Table 9.2 shows this effect for physics grades relative to grades in some other subjects.

This table shows that, on average, candidates achieved lower grades in physics than they did in their other A levels, often as much as one grade. Similar tables can be derived for other subjects. These type of studies raise questions about what information these results provide. For example, should all subjects have the same difficulty in terms of the grades awarded, or should we reduce the inherent difficulty of the syllabuses in the apparently more difficult subjects of physics and chemistry. This technique of comparing pairs of subjects to study difficulty is not without its critics, and Newton (1997a) argues that since the sample pairs are self-selecting, it can be that one group is inherently more able than another, and this compounds the problem. It

Table 9.2   Subject pair analysis for A level physics and another subject

A level physics compared with	Grade difference
Biology	+0.96
Mathematics	+0.15
Chemistry	+0.07
Business studies	+1.10
Sociology	+1.64
History	+1.01
General studies	+1.50

may be the case with the physics entry, which is often derived from the upper end of the ability spectrum and then paired with the whole spectrum of ability in another subject. Essentially, further work needs to be done here, which looks at the quality of the entry before making causal statements about difficulty. The simple lesson here is that any conclusions from the analysis of summative data *must*, like any other data, be examined with respect to the *underlying assumptions* and not taken at face value.

Standards

In many countries, making value judgements on the standards of achievement has become a national obsession, particularly at the times of the reporting of public examination results. Useful research evidence is difficult to obtain and Newton (1997b) suggests that such judgements are impossible, in so far that syllabuses, examination styles, and teaching, change over time. Consequently, studies which attempt to show changes over time must be viewed with caution. Two examples are provided to indicate some possible inferences.

The first example uses recent data to show how the grading of the UK 16+ GCSE Science Double Award examinations has changed over time (Table 9.3).

This table shows that there have been increases in the proportion obtaining grades A and B and this is particularly pronounced for grade B, although this appears to have been re-aligned to former percentages in 1998. This general increase appears to be at the expense of grades E to U where there is a reduction in the proportion awarded. Grades C and D have remained relatively static. By looking at these patterns, we could make a sweeping statement such as – standards are declining because more students are passing with higher grades, or alternatively – standards are rising because teaching has improved, students are better prepared and achieve higher grades. In both cases we do not know where the 'truth' lies because the evidence base is unsound. For example, the syllabuses have changed and the National Curriculum (science) has been introduced, so it is questionable whether like is

Table 9.3   Percentage of candidates achieving grades in Science (Double Award) from 1989 to 1998

Grade	A	B	C	D	E	F	G	U
1989	6.4	11.4	22.3	19.6	17.9	12.3	6.5	3.5
1991	9.4	12.2	21.0	20.0	17.4	11.9	5.5	2.5
1993	10.3	12.6	22.9	19.3	16.5	11.3	5.2	2.0
1995	10.6	18.3	20.8	21.5	16.2	8.7	2.6	1.3
1997	10.5	17.6	20.4	21.8	16.2	9.1	2.8	1.7
1998	10.9	12.6	25.2	21.5	15.4	8.7	3.4	2.3

Source: SCAA (1996a).

being compared with like. Further work needs to be done here before any claims can be justified.

The second example takes another stance (SCAA 1996b, 1996c) in which syllabuses, examination papers and candidates' scripts in various subjects were looked at in detail over a ten-year period. Chemistry examinations at 16+ and 18+ were chosen for the sciences. The detailed analysis that was carried out indicated that judgements of decline or improvement were difficult. For example, the results from the 18+ chemistry study showed that some changes have taken place. There have been reductions in the mathematical and the inorganic chemistry demands of syllabuses. There has been greater use of structured questions, and candidates are now expected to use more skills such as interpreting data rather than recalling knowledge. There has been a simultaneous decline in performance in inorganic chemistry and in the use of symbolic representations of equations. A more recent report by the Royal Society of Chemistry (1998) found that standards had declined by the equivalent of two A level grades between 1989 and 1996. The findings are based on the results of a chemistry test given to new undergraduates, which were then correlated with their A level grade. For example, a grade A candidate in 1989 would have scored 82 on the test but would score 75 in 1996. A similar decline is shown for other grades. However, the fundamental problem of comparability of standards still remains because, if the syllabus changes or the pedagogy has changed during the interim, candidates will have been exposed to a different educational experience during the period under investigation, which will lead to a different performance on such an unchanged chemistry test.

National views by other means

Some countries attempt to monitor standards through systematic research. For instance, in the UK, the Assessment of Performance Unit (APU) was set up in 1975 to promote the development of methods of assessing and monitoring of the achievement in schools and to identify any under-achievement

(Black 1990). Data were gathered annually from 1980 to 1984 with the focus on pupils aged 11, 13, and 15 (Johnson 1989). The surveys used were extensive, typically 12,000–16,000 pupils and 300–600 schools. The assessment framework for science that was originally used was to assess students abilities to:

- use graphical and symbolic representations;
- use apparatus and measuring instruments;
- make observations;
- interpret and apply scientific knowledge;
- plan investigations;
- perform investigations.

There was wide use of pencil-and-paper tests, made up of questions derived from a bank containing many hundreds of pre-tested questions, each question carefully targeted to a specific science area and having a context defined in terms of everyday or scientific. Answers were analysed to see what type of response the pupils had made. Pupils' scientific achievement was found to vary considerably with the context of the question, with everyday contexts usually performing better. As a result of this work, the context of the question is carefully considered and scrutinized in questions for national examinations. In addition, issues of validity and reliability, which were always in the forefront of the APU study, showed that, if the assessment was to have content validity, then a hands-on practical assessment was essential to reflect the nature of science. Consequently, a novel feature was the testing of practical skills and the looking at performance on planning and carrying out investigations on a national scale. This work effectively raised the status of the necessity to assess practical work at all ages for the next decade.

Perhaps one of its greatest achievements was to raise awareness of assessment issues and provide a legacy which was crucial to the development of criterion-referenced assessment frameworks such as 'The Graded Assessment in Science Project' (GASP) (Swain 1989), where pupils were monitored on all aspects of their science during secondary schooling in order to achieve a cumulative and graded profile, rather than a norm-referenced method of assessment which has no points of reference for its final judgements on pupils. As such, this work was influential in developing the first assessment programme for the UK National Curriculum (Taylor 1990; Swain 1991a, 1991b), and its effects still pervade current work on assessment.

International views: TIMSS surveys

The last decade has seen assessment data move towards providing information for policy-makers. Individual countries may have their own idiosyncratic systems of education, but politicians are placing an increasing emphasis on how their systems perform in comparison to the rest of the world. Whether such national indicators have a direct and immediate effect on policy is uncertain; nevertheless, countries will always prefer to be in the upper quartile rather than in another, which might indicate that they are far from their stated goals.

The third and most recent international study for science and mathematics was undertaken by the International Association for the Evaluation of Educational Achievement in the early 1990s. Known as the Third International Mathematics and Science Study (TIMSS), it was the largest and most comprehensive international study ever undertaken of educational achievement, and involved nearly half a million 4, 8 and 12 grade students in nearly 50 countries and 15,000 schools. Its main objectives were to compare and analyse curricula, teaching practices and student achievement in science and mathematics in the participating countries; to enable countries to determine whether they were internationally competitive; to examine the variety of best practices in successful schools and finally to establish world-wide benchmarks for setting and evaluating goals in mathematics and science (Murphy 1996). Its methodology was very broad as, not only were students tested in the conventional way, but teaching practices, the role of the curriculum in teaching and learning, textbooks, homework, and student attitudes, were all also examined. It is from such studies that we can gain an insight into how different societies and cultures with different national educational policies can influence the achievement of students within. Only the formal testing, which took place in grade 8, will be discussed in this section.

International achievement at grade 8

In these assessments, five scientific domains were tested by means of pencil and paper with each domain containing a number of items (135 in total across all domains). They were – earth science, life science, physics, chemistry, environmental science and the nature of science. The scores from these were combined to produce an overall science score and then arranged in a hierarchy. There was also a less extensive study of the assessment of the practical domain, which was carried out by means of performance assessments on specific practical tasks, and this area will be looked at later. The results of the written components testing the five domains are given in Table 9.4.

Singapore together with the Czech Republic, Japan and South Korea were the top performing countries, and Columbia, Kuwait and South Africa the lowest. England and Wales, although not first, were in the upper quartile. Not shown in the table are Ireland which scored a total of 538 and Scotland a total of 517.

Further information in the study reveals that there are gender differences in performance, and in most countries boys had significantly higher achievement than girls. This was mainly due to the higher performance of boys in the earth science, chemistry and physics areas. Perhaps more surprising from the table is the high international performance of all countries in the life sciences area and the lower performances on chemistry, environmental issues and the nature of science. This, of course, not only raises issues about the

Table 9.4   Scores from the TIMSS grade 8 test survey based on overall mean score for the science assessment

	Overall mean score	All science content areas	Earth science	Life science	Physics	Chemistry	Environ-mental science and Nature of science
Singapore	607	70	65	72	69	69	74
Czech Republic	574	64	63	69	64	60	59
South Korea	565	66	63	70	65	63	64
Netherlands	560	62	61	67	63	52	65
Austria	558	61	62	65	62	58	55
England and Wales	552	61	59	64	62	55	65
Belgium	550	60	62	64	61	51	58
Australia	545	60	57	63	60	54	62
Average all countries	516	56	55	59	55	51	53
Lowest (South Africa)	326	27	26	27	27	26	26

Source: Beaton *et al.* (1996).

international performance in these areas, but also about the validity of the questions used to assess this performance. However, such data do provide countries with the opportunities to re-examine their curricula and pedagogy in order to identify possible weaknesses and to implement strategies to correct these in the future, and hence use the information obtained in a formative way.

The performance assessments

Science is regarded as a practical subject and to ignore this domain, as the APU work showed, would be automatically to invalidate the overall assessment. Only 19 countries took part in this component (IEA(TIMSS) 1996/1997), and so the overall validity of the assessment in all the countries must be questioned. In these performance assessments five practical tasks ('pulse', 'magnets', 'batteries', 'rubber band' and 'solutions') were used to assess the practical domain. Each task had specific performance criteria; so, for example, in the 'pulse' task, where students had to look at changes of rate on exercising, the presentation and quality of the data from the measured pulse, a description of the trend due to the increasing exercise and an explanation of the results were all assessed.

The results show that the performance of English and Welsh children was second only to Singapore, and better in this domain than the written component. If the standard errors are taken into account, then the two results are comparable. One likely reason for the high performance here must originate from the students' exposure to the investigative practical work in the Science National Curriculum.

A fuller analysis shows that there is variation in performance across each country on each task. For example, Columbia, the lowest scoring country, produced one of the highest scores for the 'magnets' task. There was also variation between tasks; 'pulse', for example, was the most difficult task and 'magnets' was the easiest. This implies that students should be given experience of investigational practical work in a variety of scientific domains rather than in a single domain.

Simple tables of results can hide much of the richness of the data collected. For example, despite overall similarities, there are wide differences in the performance on specific skills, such as collection and presentation of results, in the tasks between Singapore and England and Wales. Reasons for these differences are not clear and need further research. However, again, the data do highlight information which would have otherwise have lain dormant, and which might be used in a formative way to the benefit of future students.

Conclusions

The previous sections lead us to three important interrelated issues about summative assessment. The first is the concept of power within assessment and how it can determines the type (formative or summative); the second relates to the use and purpose of summative assessment, and the third seeks a new definition for summative assessment.

Acknowledging summative assessment as an instrument of power

If we were to formulate a theory of assessment, then the concept of power would need to be introduced. Assessment has always been an instrument of power. It provides a way for controlling students, people, organizations and systems; it identifies progress, it puts students into hierarchies, it can be used to select and decide futures, and it can be used to make decisions. It is who controls this power, and what is done with the information derived from it, that is important and that also determines if the assessment is summative or formative. Most commonly, the locus of power for summative assessment resides with an organization, such as with an examination board or with Government legislation. It tends to serve the interests of the 'powerful' and not the interests of the students. Unless the interests of the students are incorporated within the context of summative assessment with a *post hoc* evaluation, then the clear distinction between formative and summative assessment will remain. It is only when we learn to use the results of summative assessment more effectively as an automatic post-summative reflection that it will become a formative exercise. Then intrinsic improvements in the learning experience offered to future students will take place. Then the classic distinctions between summative and formative assessment might blur.

The need to clarify purpose and use in summative assessment

Those constructing summative assessments always look at the technical issues of the reliability in terms of the consistency of results, and the validity, in terms of the credibility of the assessments that are given and the results produced. These are particularly important with respect to the setting of national and international tests and public examinations. However, there are also socio-economic issues associated with summative assessment such as cost, uses, time, effort, impact on staff and students, and benefits to society. This is a newer area in assessment research and is little explored or made explicit at present. However, these technical and socio-economic issues are beginning to merge and Messick's (1989) ideas on re-defining validity are influential here. He introduced the notion of consequential validity, which links both the purpose and the use of the assessment. These are both elements in summative assessment, but both lie outside the control of the individual student. The purposes are paramount, and future policies on summative assessment should ensure that clear aims are given for its use, whether it be for the evaluation of programmes and schemes of work, teachers in science departments, student performance at various ages, or to help schools and parents with decision-making. It is only then, when the intentions are clear, that summative data will be used more effectively for the benefit of future students.

Towards a newer definition of summative assessment

The existing definitions of summative assessment seem to be limited and too often fail to utilize the potential in the data generated. Summative data should lead, almost automatically, to questions, to analysis, and hopefully to answers, and so provide the essential ingredient in the feedback loop of teaching, learning and assessment. So far, summative assessment data have been under-used by teachers. Consequentially, the definition of summative assessment needs to be re-examined in the light of its potential. A newer definition might be that summative assessment is one which has a pre-defined purpose and will produce data on an individual, or individuals, at some point in time, which can then be used both to inform and to enhance the teaching and learning of future cohorts of students. The implication of this definition is that summative assessment should always have some formative function. The *information* component within this definition is well established – reports are produced for students, parents, teachers, school governors, LEAs, and government. The actions which follow are less clear, and decisions about educational interventions, teaching and course modifications, resource allocations, policy and research for the intrinsic improvement of the educational system are often ignored. The intrinsic value of summative assessment has yet to be realized.

References

Apple, M.W. (1992) Educational reform and educational crisis, *Journal of Research in Science Teaching*, 29(8): 779–89.

Ball, S. (1990) *Markets, Morality, and Equality in Education*. London: Tufnell Press.

Beaton, A., Martin, M., Mullis, I., Gonzales, E., Smith, T. and Kelly, D. (1996) *Science Achievement in the Middle School Years: IEA's Third Mathematics and Science Study (TIMSS)*. Chesnut Hill: Boston College.

Berk, R.A. (1980) A consumers' guide to criterion-referenced test reliability, *Journal of Educational Measurement*, 17(4): 323–49.

Black, P.J. (1990) APU science – the past and the future, *School Science Review*, 72(258): 13–28.

*Black, P.J. (1993) Formative and summative assessment by teachers, *Studies in Science Education*, 21: 49–97.

Black, P.J. (1994) Performance assessment and accountability: the experience in England and Wales, *Educational Evaluation and Policy Analysis*, 16(2): 191–203.

*Black, P.J. (1998a) *Testing: Friend or Foe? Theory and Practice of Assessment and Testing*. London: Falmer Press.

Black, P. (1998b) Learning, league tables and national assessment: opportunity lost or hope deferred? *Oxford Review of Education*, 24(1): 57–68.

Bloom, B.S., Hastings J.T. and Madhaus, G.F. (eds) (1971) *Handbook on the Formative and Summative Evaluation of Student Learning*. New York: McGraw-Hill.

Borrows, P. (1997) Analysing examination statistics, *School Science Review*, 79(286): 47–9.

Butler, J. (1995) Teachers judging standards in senior science subjects: fifteen years of the Queensland experiment, *Studies in Science Education*, 26: 135–57.

*Butterfield, S. (1995) *Educational Objectives and National Assessment*. Buckingham: Open University Press.

Conner, C. and James, M. (1996) The mediating role of LEAs in the interpretation of government assessment policy at school level in England, *The Curriculum Journal*, 7(2): 153–66.

DES (1988) *National Curriculum: Task Group on Assessment and Testing: A Report, Department of Education and Science and Welsh Office*. London: HMSO.

DfEE (1995) *Value Added in Education*. London: HMSO.

DfEE/Ofsted (1996) *Setting Targets to Raise Standards: A Survey of Good Practice*. London: DfEE.

Fitz-Gibbon, C.T. and Vincent, L. (1994) *Candidates' Performance in Public Examinations in Mathematics and Science*. A report for SCAA. Newcastle: University of Newcastle upon Tyne.

Gibson, A. and Asthana, A. (1998) School, pupils and examination results: contextualising school performance, *British Educational Research Journal*, 24(3): 269–82.

IEA (TIMSS) (1996/7) *Science Achievement Reports for the Third International Mathematics and Science Study*. Boston Hill: Centre for the Study of Testing, Evaluation and Educational Policy.

IEA (TIMSS) Harmon, M., Smith, T., Martin, M. *et al.* (1997) *Performance Assessment in IEA's Third International Mathematics and Science Study*. Boston Hill: Centre for the Study of Testing, Evaluation and Educational Policy.

Jesson, D. (1997) *Value Added Measures of School GCSE Performance*. London: DfEE.

Johnson, S. (1989) *National Assessment: The APU Science Approach*. London: HMSO.

Kempa, R.F. and Odiaga, J.L. (1984) Criterion-referenced interpretation of examination grades, *Educational Research*, 26(1): 56–64.

Lang, H.G. (1982) Criterion-referenced tests in science: an investigation of reliability, validity, and standards-setting, *Journal of Research in Science Teaching*, 19(8): 665–74.

Messick, S. (1989) Validity, in R.L. Linn (ed.) *Educational Measurement*, 3rd edn. New York and London: Macmillan and American Council on Education.

Murphy, P. (1996) The IEA assessment of science achievement, *Assessment in Education: Principles, Policy and Practice*, 3(2): 129–41.

Murphy, R. (1997) Drawing outrageous conclusions from national assessment results: where will it all end? *British Journal of Curriculum and Assessment*, 7(2): 32–4.

Newton, P.E. (1997a) Measuring comparability of standards between subjects: why our statistical techniques do not make the grade, *British Educational Research Journal*, 23(4): 433–49.

Newton, P. (1997b) Examining standards over time, *Research Papers in Education*, 12(3): 227–48.

Popham, J.W. (1978) *Criterion Referenced Measurement*. Englewood Cliffs, NJ: Prentice-Hall.

QCA (1998) *Standards at Key Stage 3: Science A Report on the 1997 National Curriculum Assessments for 14 year olds*. London: Qualifications and Curriculum Authority.

Royal Society of Chemistry (1998) *Research in Assessment XIII (An Updated Report on the Skills Test Survey of Chemistry Degree Course Entrants)*. London: Royal Society of Chemistry.

SCAA (1996a) *GCSE Results Analysis*. London: School Curriculum and Assessment Authority.

SCAA (1996b) *GCE Results Analysis*. London: School Curriculum and Assessment Authority.

SCAA (1996c) *Standards in Public Examinations 1975 to 1995*. London: School Curriculum and Assessment Authority.

SCAA (1997) *Making Effective Use of Key Stage 2 Assessments*. London: School Curriculum and Assessment Authority.

Strand, S. (1998) A 'value added' analysis of the 1996 primary school performance tables, *Educational Research*, 40(2): 123–37.

Swain, J.R.L. (1989) The development of a framework for the assessment of process skills in a Graded Assessments in Science Project, *International Journal of Science Education*, 11(3): 251–9.

Swain, J.R.L. (1991a) Standard assessment tasks in science at Key Stage 3: initial development to the 1990 trial, *British Journal of Curriculum and Assessment*, 1(2): 26–8.

Swain, J.R.L. (1991b) Standard assessment tasks in science at Key Stage 3: the 1991 pilot, *British Journal of Curriculum and Assessment*, 2(1): 19–20.

Swain, J.R.L. (1992) Trialling and piloting KS3 science SATs, *School Science Review*, 74(267): 115–20.

Swain, J.R.L. (1996) The impact and effect of Key stage 3 science tests, *School Science Review*, 78(283): 79–90.

Taylor, R.M. (1990) The National Curriculum: a study to compare levels of attainment with data from APU science surveys (1980–4), *School Science Review*, 72(258): 31–7.

*Taylor Fitz-Gibbon, C. (1996) *Monitoring Education: Indicators, Quality and Effectiveness*. London: Cassell.

Wiliam, D. (1993) Validity, dependability and reliability in National Curriculum Assessment, *The Curriculum Journal*, 4(3): 335–350.

Wiliam, D. and Black, P. (1996) Meanings and consequences: a basis for distinguishing formative and summative functions of assessment, *British Educational Research Journal*, 22(5): 537–48.

10 Science teaching and the development of intelligence

Philip Adey

In the Cognitive Acceleration through Science Education (CASE) project we have demonstrated that science teaching can be used to raise students' intelligence. Such a bold claim should raise a great many questions in the minds of a healthily sceptical scientist. In this chapter I propose to address a couple of those questions, and to see what answers research can provide – and with what level of confidence. Considering the mind as an information-processing mechanism, I will briefly outline issues to do with 'paying attention' and then look more closely at the nature of intelligence, its malleability, and the nature of activities in science lessons which might promote the development of intelligence.

Pay attention!

Every waking minute of our lives, we and our students are subject to a vast input of sensory experiences. Our eyes, ears, nose, skin, and other sense organs are bombarded with stimuli. Effective learning requires us to (a) attend especially to those stimuli which are relevant to the learning task in hand, and (b) interpret and make sense of those relevant stimuli. This chapter is mostly about the second process, making sense of inputs, but we cannot ignore the first process entirely.

The business of 'paying attention' is not at all well understood. We know that it is absolutely necessary for our minds to filter out perhaps 95 per cent of the stimuli received, in order to avoid hopeless overload and confusion. We know also that self-control over attention is limited. Everybody, at some time, has found their mind wandering during even the most interesting lecture, piece of music, or play. Masters of modern media have learned the art of maintaining audience attention during a television or film presentation by rapid change of pace, story-line, or viewpoint. They know that variety plays an important role in grabbing attention. The point for us is that we soon become accustomed to routine, and that change is one of the most effective ways of maintaining attention.

Of course, other factors are important also – degree of tiredness, hunger, temperature, humidity, and carbon dioxide concentration all play their part and, perhaps most obviously, the intrinsic interest of some subjects. For example, sex is intrinsically interesting for good reasons of species survival. See Kellogg (1995) Chapter 3 for a full treatment of attention.

The information processing factor

Supposing, as a teacher, you pull out all of the stops in terms of keeping attention: you use a variety of paces, materials, and changes of attack, appeal to a variety of learning styles, using visual, verbal, and numerical material, and make it all as relevant as you can to your students' interests. On occasion, you can do all that is possible to gain and maintain attention, and yet you are conscious that not a lot of learning is taking place. Perhaps the problem lies in the other factor, the ability of your students to process the information you are providing. Michael Shayer (Shayer and Adey 1981) has described how limits on information-processing ability of students (described in terms of their stage of cognitive development) provide a powerful explanatory model for the difficulty they encounter with some concepts in science.

There is a substantial literature on the mind as an information-processor (Case 1975; Pascual-Leone 1976; Johnstone and El-Banna 1986; Liben 1987; Voss 1989; McGuinness 1990). At their simplest, Information Processing (IP) theories treat the mind like a computer. There are inputs, processing, and outputs. The power of the mind/computer is determined by the power of the processor, where *power* here has the same basic meaning as in physics: rate of doing work. However, increases in power of a machine may be accomplished by increasing the speed with which it does the same thing (lift a bucket, or whatever). We will see that increases in power of the mind necessarily entail a change in the quality of processing – specifically an increase in the number of bits of information that can be handled at the same time.

There is also a substantial literature on the nature of intelligence (Carroll 1993; Ceci 1993; Herrnstein and Murray 1994; Byrne 1995; Perkins 1995; Woodcock 1995; Deary and Stough 1996; Gardner *et al.* 1996; Sternberg 1998). In what follows I will explore the ways in which information-processing ideas and ideas about intelligence may be integrated.

What counts as *intelligence*?

The first activity I ask new CASE teachers to do on our Professional Development days is to think about what counts as intelligent behaviour in their students. What sort of thing does a student have to say, or write, or do, to make you say, 'that's smart'? (You might like to consider this yourself for a minute, before reading on.)

As you might expect, I get a great many answers to that question, but the most common are things like:

- can apply existing knowledge to new situations;
- makes connections between different areas of knowledge;
- sees a pattern in data;
- asks searching questions.

If you have ever looked at verbal, numerical, or pictorial intelligence tests (for example Wechsler 1958; Raven 1960; NFER-Nelson 1993), you will have noticed that many of the items demand just this type of thinking: making connections, seeing analogies or patterns, or abstracting some general pattern. So, it seems that psychologists who devise intelligence tests and teachers agree broadly about the nature of intelligence. It has something to do with making connections, connectivity.

This idea of connectivity can be unpacked at various levels, which I will describe as *macro-psychological, mind-model*, and *neuronal*. Research at each level provides us with insights, and if we ourselves have sufficient intellectual connectivity, we may be able to integrate the insights from all three levels to form a well-articulated conception of the nature of intelligence.

Macro psychological

These are the ways in which psychologists since Alfred Binet (1909) at the beginning of the twentieth century have tried to provide operational descriptions of intelligent behaviour, which can be used as a basis of assessment of intelligence. Binet's own descriptive work (well described by Perkins 1995) was taken up by the Americans and made into a quantitative measuring instrument (Thorndike *et al.* 1986), which yielded an Intelligence Quotient (IQ) for each individual that indicated how their intellectual ability compared with others of the same age. This process of turning descriptions into measurements offered great reliability and predictive validity, but it lost much of the richness of description of intelligent behaviour. An IQ score will give you a pretty good idea how an individual will fair in normal academic achievement, but it will not tell you much about why, or what you might do about it.

Meanwhile in Europe, one of Binet's students was Jean Piaget. On the basis of a lifetime of research and analysis, starting with observations of his own children, Piaget (1950) developed rich and detailed descriptions of the development of intelligence. With Bärbel Inhelder (Inhelder and Piaget 1958), he described the highest level of intellectual performance as *formal operations* and characterized it as the ability to, for example:

- hold many variables in mind at once and operate upon them;
- use abstract ideas in conjunction with one another;
- see actual events as a sub-set of many possible events.

Here we see the connectivity idea very clearly. If you are going to see connections between science concepts (say to see the relationship of respiration to photosynthesis), you need to be able to hold in mind at once the important characteristics of each and also be able to compare them. This is multi-variable thinking. Many models, which are central to scientific understanding, such as current flow, or kinetic theory, are abstractions. You cannot handle them physically, but can only come to understand them through a thorough familiarity with their characteristics and applications in the real world. The investigation of cause and effect, the design of experiments, requires the ability first of all to hold all possibilities in mind, and then systematically to eliminate possible causes one by one.

Robert Sternberg is probably the leading contemporary cognitive psychologist who writes extensively about intelligence (Sternberg 1996 is a good starting place). He describes a *triarchic* theory, which encompasses contextual, experiential, and componential sub-theories of intelligence (Sternberg 1985). Of these, it is the *componential* theory which will provide us with an important *extra* to the simple information-processing model described above. That extra is what Sternberg describes as a metacomponent, higher-order executive processes used in planning, monitoring, and decision-making by an individual engaged with a problem. We will see that this proves to be an important factor in helping students to develop their intelligence.

Mind models

Cognitive psychologists have proposed a variety of mind models, but all of them include the ideas of a *long-term memory*, a *working memory*, and some sort of *executive control* or *central processing mechanism* (Baddeley 1990; Anderson 1992). Evidence for the existence of long-term memory is clear (Baddeley 1990) and, in fact, shows that apparent failure of memory is not due to material being lost from the store, but to temporary or longer-term failure of the central processing mechanism to recover information that is there. This may be illustrated by the common phenomenon of an elderly person who may forget what happened yesterday, but whose memory of their childhood becomes much clearer.

On this most basic model, the sense organs detect external stimuli (patterns of light, sound, and so on) and convert them into electrochemical messages in neurons. The working memory receives these signals, and also may receive information from long-term memory. Information from various sources, internal and external, may be compared and combined in working memory, creating new syntheses or simply *recognizing* an external input as corresponding to a known configuration from long-term memory. On the basis of such comparisons, combinations, and recognitions, working memory may produce outputs to activate motor nerves or for storage in long-term memory. The executive processor controls these various inputs and outputs.

There are a number of elaborations and refinements of this basic model, which will be relevant to an understanding of how the right kind of science teaching can enhance intelligence. One such elaboration (Pascual-Leone 1976; 1984) posits that working memory actually has a limited number of *slots* to hold bits of information, and that the number of slots grows from just two at birth to a maximum of seven in mature intelligent adults. One way of testing working memory span (Towse *et al.* 1998) is to present a series of very simple sums (for example 5 + 1 = . . .) and ask the person to remember the answer to each. After four, five, or six sums the person is asked to recite all the answers so far. Holding more than seven proves to be very difficult, even for intelligent adults. (If you just give someone a string of numbers to remember, they can recite the string to themselves, or find a pattern to help them *chunk* the number and so lower the working memory demand. Interspersing the sums, trivial in themselves, is intended to deny this possibility.)

This view of the growth of working memory provides a useful explanation of the development of intelligence, and ties in with our picture of intelligence as concerned with multi-variable thinking and the ability to hold a number of ideas in mind at once. Growth in power of the mind requires a development of parallel processing ability. This is necessary to meet our connectivity requirement for intelligence, since in order to see the connection between two things, both have to be held in mind at once. At an elementary level, when water is poured from a squat beaker into a tall beaker, an average 5-year-old will believe that the amount of water has changed (Piaget and Inhelder 1974). This can be explained if one supposes that limitations in the child's working memory prevent him/her from considering, *at the same time*, the height of the water and the breadth of the water, and so constructing a compensation explanation for the amount of water remaining constant. At a senior secondary science level, the ability really to understand the law of moments (as opposed to memorizing the algorithm and slotting in the numbers) requires that the student hold in mind two masses, two distances, and whether the system is in balance or not. That is five bits of independent information, quite a demand on working memory.

A refinement of the model due to David Ausubel (1968) describes knowledge held in long-term memory in the form of networks of interconnected bits of information. The more sophisticated is the elaboration of these networks, the better prepared is the whole mind to assimilate new information about a particular topic, since the working memory and processor have a richer source of existing knowledge to which to relate new inputs. This explains why it is easier for us to understand and to learn new things in a field with which we are already familiar, and why starting to learn in a completely new field is so difficult.

A further elaboration of our mind model should be mentioned for the sake of completeness, although it is not of direct relevance to our present purpose. This is the addition of a small number (from one to four) of specific *modules* responsible, for instance, for language development (Chomsky 1986). Such modules are supposed to be inherent and a product of human evolution. The evidence for the existence of a language module lies in the incredible facility with which infants learn a language – far faster and in far greater elaboration than could ever be explained by a simple association learning process. At four days old, an infant can distinguish language from other sounds, and at four weeks old can distinguish his mother tongue from a foreign language. This is strong argument for the human brain being hard-wired with a predisposition to learn language.

Neuronal

Cognitive neurophysiology is making enormous strides in developing our understanding of the structure and function of the brain. Aspects of the localization of brain functions have been known for many years from evidence of patients with brain lesions, but a range of new techniques have led to an explosion in knowledge of brain architecture (Greenfield 1998; Johnson 1997).

- Modern microscopy has allowed anatomists to look directly at neurons (nerve cells) in the cortex.
- Positron Emission Topography (PET) is a technique in which a subject is injected with glucose containing radioactively labelled molecules. On the assumption that glucose metabolism in a brain locality indicates activity at that site, three-dimensional positron scanning equipment can follow the localization of brain activity.
- Event-Related Potentials (ERP) are tiny changes in electrical activity which can be detected with electrodes attached to the scalp.
- Magnetic Resonance Imaging (MRI) allows blood flow rates (assumed to be associated with activity) in parts of the brain to be measured and, again, such activity can be tracked as the individual engages in different types of cognitive activity.

Although none of these methods is cheap, comfortable for the subject, or very easy to interpret, they do allow for genuine experimentation, of the form, 'Let's see what would happen if . . .'. For our present purposes, just a couple of the results of this exciting new research area are relevant.

1 Rats brought up in a stimulating environment, with many colours, puzzles, and toys, develop far more complex neural networks than initially matched rats brought up in normal laboratory cages (Greenhough *et al.* 1987). This is the most direct evidence that the brain develops physiologically in response to external stimulus.
    There is actually a strong argument that it could not be otherwise: we know that behaviour can be changed by experience. Behaviour is a

response of the brain, so behaviour change implies a change in the brain, and the effect of experience on behaviour shows that experience must change the brain. These brain changes must be physiological changes – the only alternative is a dualist model, which includes some aspect of mind (spirit? soul?) that is independent of a physical brain, a proposal unlikely to appeal to those of a scientific turn of mind.

2 The brain first starts to form in the human embryo about 12 days after conception. From then until birth, there is a massive growth in the number of neurons in the brain (around 250,000 per minute before birth). After birth, growth in the number of neurons virtually ceases, but there is a series of phases of growth of dendritic connections (complexification of the neuronal network) and decay of (apparently) unused connections. These growth and decay cycles occur in different locations in the brain at different ages, but are not completed until some final organization in the forebrain in early adolescence. Johnson (1997) provides a comprehensive account of brain development.

We see here the intimate interaction of human evolution, individual heredity, and environment. The course that an individual's cognitive development actually takes will be determined by a continual interaction between evolutionary predispositions, specific genetic make-up, and environmental influences.

The neuronal level of considering the development of intelligence shows up the inadequacy of the 'brain-as-computer' idea floated in the introduction to this section. The brain is a living organism, and the process by which it develops depends at least as much on its ability to select and reinforce useful neuronal pathways as on any pre-programming. This is a position strongly supported by the work of Edelman (1987), and predicted many years ago (and in surprising detail) by Hebb (1949). The importance ascribed to environment in the development of intelligence places an awesome reponsibility on us as teachers (and as parents) as we shape the learning environment to influence the cognitive development of children in our charge.

Just one intelligence?

I have avoided until now the vexed question of whether there is just one, or many types of intelligence. The answer is: there is one, and there are many.

Howard Gardner (1993; Gardner *et al.* 1996) has proposed that there are at least seven independent types of intelligence. He has named these numerical, spatial, verbal, kinaesthetic, musical, interpersonal, intrapersonal, and naturalist. The theory of multiple intelligences allows one to suppose that a child may be, 'brilliant at maths but useless at English' and, by extension, that everyone has some sort of talent that can be developed. I believe it is only because this is an idea very comforting to liberal-thinking teachers that the idea of multiple intelligences has been so widely accepted. In a book devoted to 'what research says' to the science teacher, we are bound to look critically at evidence, and the evidence of the existence for abilities in different intellectual areas, which are quite independent of one another is not good. Certainly, intelligence is not a monolithic unidimensional ability, which allows us with one IQ number to define an individual fully. But, all measures of different aspects of intellectual ability correlate with one another (Anderson 1992). By any sensible statistical analysis (Carroll 1993), the most reasonable way to explain the vast amount of data which comes from multiple testing of individuals is in terms of (1) a general underlying intelligence ('g') plus (2) a number of specific abilities such as verbal, numerical, and spatial. Any intellectual behaviour is then a product of a general processing ability and a number of specific abilities. In terms of the mind models discussed above, the general processor may be a function of the executive control of working memory, and the specific abilities rest either in specialist modules such as Chomsky's General Language Processor, referred to earlier, or in particular networks of concepts established early in life in long-term memory.

How all this relates to science teaching

It is about time I turned from this instant introduction to cognitive psychology and addressed the question of its relevance to science teaching in schools. I have described high-level thinking as requiring an individual to be able to hold in working memory, at one time, four, five, or more bits of information, and to be able to relate them to one another. This *multi-parallel processing* is what Piaget describes as formal operations. If you look at Science in the National Curriculum, at the Level Descriptors or Programmes of Study (PoS), it is possible to use this particular notion of intelligence to assess the relative difficulty of different concepts. Here are some examples from the different attainment targets within the KS4 Programmes of Study. The final column of Table 10.1 gives an estimate of the level of thinking required in terms of Piagetian stages of cognitive development:

2A early concrete operational;
2A/2B mid concrete;
2B late concrete;
2B* concrete generalization;
3A early formal operational;
3B mature formal operational.

A tool called a Curriculum Analysis Taxonomy which allows anyone – with a bit of effort – to analyse curriculum material for level of difficulty is given in Shayer and Adey (1981), now out of print but available from the authors. A complete analysis of the 1991 National Curriculum for Science – the one with the 17 ATs – is shown in both Adey (1993b) and Adey and Shayer (1994). There are a number of implications of this type of analysis.

Table 10.1　Examples of science in the National Curriculum matched to Piagetian levels

AT and PoS	Analysis	Level
AT2, PoS 1d: how substances enter and leave cells through the cell membrane by diffusion, osmosis, and active transport	This is not only inaccessible to concrete experience because none of the entities or ideas can be seen, but it requires the distinction of characteristics of different postulated processes acting simultaneously.	3B
AT2, 2e: how breathing, including ventilation of the lungs, takes place	This can be directly modelled with concrete apparatus and feeling one's own rib and diaphragm movement. The two processes of flattening the diaphragm and expanding the rib-cage can be considered separately (serially) as additive in effect.	2B
AT3, 1h: that chemical bonding can be explained in terms of the transfer or sharing of electrons	Implicit in this is some understanding of the distinction between atoms and molecules, between elements and compounds, and of an abstract model of electron structure. The real things are not accessible to the senses, and using atomic models requires a process of keeping model and real thing in mind, seeing what is similar and what is different. As a blackboard exercise with its own internal rules, it is accessible at the early formal stage, but mature formal operations are required for a useful working understanding.	3A
AT3, 2l: an example of how a reactive metal can be extracted by electrolysis	Example of a technical process involving concrete apparatus and a series of steps, each of which can be considered one at a time – serially processed.	2B
AT4, 1g: how current varies with voltage in a range of devices . . .	If 'current' and 'voltage' are taken as numbers read from meters, then the relationships can be found as soon as true proportionality is available (early formal). But, a deeper understanding requires the distinction between two abstract constructs and their relationship in a variety of circumstances. This makes a heavy demand on working memory, especially to develop the understanding initially, before it becomes chunked and automated.	3B
AT4, 1r: that like magnetic poles repel and unlike magnetic poles attract	Accessible by direct experience, once the idea of 'poles' has been established. Poles have a simple concrete representation (pins attracted to the ends of the magnet) and attraction and repulsion can be experienced directly. The distinction of 'North' and 'South' poles requires only a distinction of two types.	2A/B

One of the most striking is the general correspondence of National Curriculum level with the level of thinking required. The curriculum was written by teachers and others with much practical experience of teaching the subject, and revisions were made in the light of feedback from teachers concerning, among other things, the relative difficulty that their students found with concepts at different levels in the curriculum. So, the actual experience of teachers trying to get across ideas gives a very good intuitive feel for the cognitive demand of those ideas. What the Piagetian (or information-processing) analysis does is to provide a theoretical explanation for teachers' intuitive logic, and this demonstrates the value of research in helping to make sense of experience and providing a predictive tool.

So, what do we do about it? Identify a problem as cognitively demanding, decide it is beyond the capability of your Year 9Z group, and retire to the prep room for a cup of coffee and wait for them to mature gently to a higher stage of thinking? Not very practical as you might have to wait a long time. In fact, without the stimulation of your teaching, you might have to wait for ever. The whole point is that, although cognitive development occurs partly in response to natural maturational processes, it also depends importantly on response to cognitive stimulation. It is the nature of that cognitive stimulation that I shall consider in the next section.

Promoting high level thinking

There are three closely related facets of learning experiences that promote cognitive development: they are challenging; they require the student to become conscious of his or her own thinking; and they make much use of language for expressing ideas as they form.

Cognitive conflict

Piaget described cognitive development as a process of the mind reaching a succession of stages of equilibrium, when the experiences of the outside world could be processed by the current state of the mental machinery. Further development occurs when events in the outside world cannot be explained by current processing ability, and a new level of thinking must be sought to handle the challenge. This leads us to propose that it is activities that students find a little bit difficult, which are most likely to promote cognitive development. Vygotsky (1978: 82) says, 'the only good learning is that which is in advance of development'. In other words, the busy-work we all occasionally engage in with our pupils, when we give them work well within their capability, may keep them quiet and give the Inspector a picture of a busy and peaceful class, but it doesn't do much for cognitive development. To encourage cognitive growth, we need to provide a modicum of *cognitive conflict* – experiences which push students to the limit of their current processing capability, and just a little beyond.

Metacognition

If students are to take control of their own learning and development (which they must do, since we cannot be there driving them along every minute of the day), then they have to become conscious of themselves as thinkers and learners (see Chapter 1). Metacognition means *thinking about your own thinking*. Although inducing metacognition in students is not at all easy, it is well worth persisting with probes of the type 'OK you've got the right answer, now explain how you got it' past the standard answer of, 'I dunno, I just did it.' Holding up for inspection the type of thinking a student has been using makes it more likely that that type of thinking can be used again. Making thinking explicit is a pre-requisite for making it generally available. The importance of metacognition is based in:

- Piaget's idea of *reflective abstraction*, which he considers to be an essential characteristic for formal thinking;
- Vygotsky's idea of *language as a mediator of learning* (putting thoughts into words so that they can be shared and inspected); and
- Sternberg's ideas of *metacomponents of intelligence* described earlier.

These are all just slightly different 'takes' on what Michael Shayer (Adey and Shayer 1994) describes as *going above*: looking back and down on a completed bit of thinking so that its features can be recognized and it can be used again. Perkins and Saloman (1989) have shown the importance of metacognition in the development of general cognitive skills, which can be transferred from one context to another, and Brown *et al.* (1983) and Sánchez (1997) have provided technical analyses of the nature of meta-cognition.

Science and cognitive stimulation

There are quite a few programmes available for use in school, designed specifically for cognitive stimulation (often described as *teaching thinking*). We describe some of them briefly in Chapter 3 of Adey and Shayer (1994) and a much fuller account and evaluation is provided in Nickerson *et al.* (1985). Here, I will just mention two programmes, which are of direct relevance to secondary science teachers.

The first is the programme designed by Rosalind Driver and her colleagues Hilary Asoko, John Leach, Philip Scott, and others at the University of Leeds (Children's Learning in Science Project 1987). A central feature of CLIS is that children are given a full opportunity to describe their explanations for common phenomena, to share them in groups and with the whole class, for example, through posters. They are invited to design critical tests of their ideas, to put them to the test of evidence. Frequently, the experiments have surprising (to the children) results, causing them to re-adjust their explanations and ideas. You can see that this programme employs all of the features

I have described above as necessary for cognitive stimulation: the cognitive conflict of the surprising results, reflection on 'what I thought then, what I think now, and why I have changed' and much use of language justifying, arguing, relating evidence to explanations, and so on.

The second is our own Cognitive Acceleration through Science Education (CASE) set of activities. We (initially Michael Shayer, then joined by myself and Carolyn Yates) developed these specifically for the purpose of raising children's information-processing capability, and we based the activities on the factors required for cognitive stimulation outlined above. There were also other principles involved in their design. One was the provision of a period of *concrete preparation*. This is the first ten minutes or so of a CASE lesson, when some of the words which are going to be used are introduced, and the nature of the problem discussed. This phase sets up the students in preparation for the surprising or difficult-to-explain event, which causes conflict. *Bridging* means the linking of the type of thinking developed in this special, CASE, lesson with other opportunities in the science curriculum or beyond where that type of thinking will be useful. Finally, because we are working in Years 7 and 8, and children aged 12–14 years are just entering the doorway to formal operational thinking, the activities are structured by the schema which Inhelder and Piaget (1958) describe as characteristic of formal operations. These schema include the control of variables, proportionality, equilibrium, probabilistic relationships, and the use of formal models. They all require multi-variable thinking, and all can be identified as underlying aspects of the National Curriculum in science, which appear from about Level 5 and 6 onwards.

Does it work?

The effects of CASE on students has been widely reported (Adey 1993a; Adey and Shayer 1994; Shayer 1996) and here I will provide just the briefest summary. In the original experiment from 1984–87, teachers in ten schools tried the materials with some classes, and identified other matched classes as control groups, just following their normal science curriculum. Students in CASE classes (a) made significantly greater gains in cognitive development over the two-year period (Years 7 and 8, or 8 and 9) of the CASE intervention; (b) one year after the intervention (end of Year 9 or 10) scored significantly higher on tests of science achievement; and (c) three or two years after the intervention, when they took their GCSEs, scored significantly higher grades in science, maths, and in English than their colleagues from control classes who had not experienced CASE activities. We consider this long-term effect, combined with the transfer of effects from a science context across the curriculum to English, as evidence that the CASE intervention has had a fundamental effect on the students' general ability to process information – their general intelligence. Here at last are the grounds on which the statement which opens this chapter are founded.

We continue to track the effect of CASE on students, and every year we scrutinize the KS3 levels and GCSE grades obtained by schools which do, and which do not, use CASE in their Years 7 and 8. Taking a value-added approach, CASE schools score significantly higher levels and grades than would be expected, on the basis of their intake levels and on the basis of the grades obtained by non-CASE schools with similar intakes.

The evidence that CASE does raise KS3 and GCSE grades is quite convincing. Nevertheless, a word of warning is in order: academic gains of CASE students do not provide unambiguous support for the *mechanism* of cognitive stimulation described in this chapter. Alternative mechanisms which focus more on students' motivation (Leo and Galloway 1995) are plausible, and, although I think they are probably wrong, they do deserve to be researched properly (McLellan 1999).

Ups and downs

Teaching for cognitive stimulation has some obvious advantages, but it also entails some difficulties and problems.

The up side is the improvement shown by students in deep-level processing, which comes about from improved ability to hold in mind (working memory) many variables at once. Such multi-variable processing allows an individual to evaluate evidence against an initial belief or hypothesis, holding both the pre-conception and the evidence in mind at once. The genuine development of more sophisticated concepts in science requires such multi-variable thinking. In general, higher level thinking allows students to derive far more benefit, in terms of efficient learning, from any good instruction.

But, these benefits are purchased at a cost. An apparently obvious one is time. Interventions designed for cognitive stimulation may occupy 20 per cent of the time allocated to the science curriculum, and given the perception that the curriculum is already overcrowded, it seems reasonable to ask where this time is going to come from. In practice, the situation is nothing like as bad as it seems. For one thing, many of the process objectives in the curriculum are addressed directly by the thinking intervention. For another, after one year of work on improving thinking, students are able to understand the regular content-orientated instruction so much better that they make far better use of the time available to them.

Time spent in Year 7 and 8 focusing on the development of children's intellectual ability pays off – with great capital growth and interest – in subsequent years. If you take metaphors such as *delivering* or *covering* the curriculum as indicating a reasonable model of transmission of information, then you may find this hard to believe.

A more serious problem is the shift in pedagogy which is required for effective cognitive stimulation. Teaching for cognitive acceleration is different in many ways from normal high-quality instruction for science concept development, and teachers often need plenty of time to meet some of the underlying theory, to become familiar with the activities, and above all to practise the new skills with the support of a mentor or coach. By far the best way to approach this change in skills is by participation of a whole science department together, mutually supporting one another and calling in some outside assistance as necessary (Joyce and Showers 1988; Fullan and Stiegelbauer 1991; Adey *et al.* 1999).

Conclusion

In the introduction to this chapter I showed that psychological accounts of cognitive development, mind-models, and cognitive neuroscience are all optimistic about the influence of the environment on the development of intelligence. I hope I have shown how that optimism can be realized by science teachers, working within a broad context of the science curriculum. From the specific example of CASE, we have seen that time taken out of *delivering content* to focus on challenging students' thinking does pay off, quite quickly, in terms of academic achievement. It certainly requires something of a paradigm-shift in the mind-set of many teachers, from an emphasis on 'covering' content to a focus on the quality of dialogue that takes place between teacher and student and between student and student – recognizing puzzlement, uncertainty, and admission of confusion, as far more hopeful pathways to ultimate academic achievement than a neatly filled book of notes.

References

Adey, P.S. (1993a) Cognitive acceleration: science and other entrances to formal operations. Conference paper, Aix-en-Provence meeting of the 5th European Conference for Research in Learning and Instruction.

Adey, P.S. (1993b) *The King's-BP CASE INSET Pack*. London: BP Educational Services.

*Adey, P.S. and Shayer, M. (1994) *Really Raising Standards: Cognitive Intervention and Academic Achievement*. London: Routledge.

Adey, P., Bailey, M., Edwards, J. and Michael, N. (1999) An archeology of a school improvement program – what is left after four years? Conference paper, Montreal meeting of the American Educational Research Association.

*Anderson, M. (1992) *Intelligence and Development: A Cognitive Theory*. London: Blackwell.

Ausubel, D. (1968) *Educational Psychology: A Cognitive View*. New York: Holt Rinehart and Winston.

Baddeley, A. (1990) *Human Memory: Theory and Practice*. London: Lawrence Erlbaum.

Binet, A. (1909) *Les idées moderne sur les enfants*. Paris: Ernest Flammarion.

Brown, A.L., Bransford, J.D., Ferrara, R.A. and Campione, J.C. (1983) Learning, remembering and understanding, in P.H. Mussen (ed.) *Handbook of Child Psychology*. New York: John Wiley.

Byrne, R. (1995) *The Thinking Ape: Evolutionary Origins of Intelligence*. Oxford: Oxford University Press.

*Carroll, J.B. (1993) *Human Cognitive Abilities*. Cambridge: Cambridge University Press.

Case, R. (1975) Gearing the demands of instruction to the developmental capacities of the learner, *Review of Educational Research*, 45: 59–67.

Ceci, S.J. (1993) Contextual trends in intellectual development, *Developmental Review*, 13: 403–5.

Children's Learning in Science Project (1987) *CLIS in the Classroom*. Leeds: University of Leeds Centre for Studies in Science and Maths Education.

Chomsky, N. (1986) *Knowledge of Language, Its Nature, Origin and Use*. Westport, CN: Praeger.

Deary, I.J. and Stough, C. (1996) Intelligence and inspection time, *American Psychologist*, 51(6): 599–608.

Edelman, G.M. (1987) *Neural Darwinism: The Theory of Neuronal Group Selection*. New York: Basic Books.

Fullan, M.G. and Stiegelbauer, S. (1991) *The New Meaning of Educational Change*. London: Cassell.

Gardner, H. (1993) *Frames of Mind*, 2nd edn. New York: Basic Books.

Gardner, H., Kornhaber, M. and Wake, W. (1996) *Intelligence: Multiple Perspectives*. Fort Worth, TX: Harcourt Brace.

*Greenfield, S. (1998) *The Human Brain. A Guided Tour*. London: Phoenix.

Greenhough, W.T., Black, J.E. and Wallace, C.S. (1987) Experience and brain development, *Child Development*, 58: 539–59.

Hebb, D.O. (1949) *The Organization of Behaviour*. New York: John Wiley.

Herrnstein, R. and Murray, C. (1994) *The Bell Curve: Intelligence and Class Structure in American Life*. New York: Free Press.

*Inhelder, B. and Piaget, J. (1958) *The Growth of Logical Thinking*. London: Routledge Kegan Paul.

*Johnson, M.H. (1997) *Developmental Cognitive Neuroscience*. Oxford: Blackwell.

Johnstone, A.H. and El-Banna, H. (1986) Capacities, demands, and processes – a predictive model for science education, *Education in Chemistry*, 23(3): 80–4.

Joyce, B. and Showers, B. (1988) *Student Achievement through Staff Development*. New York: Longman.

Kellogg, R.T. (1995) *Cognitive Psychology*. London: Sage.

Leo, E.L. and Galloway, D. (1995) Conceptual links between Cognitive Acceleration through Science Education and motivational style: a critique of Shayer and Adey, *International Journal of Science Education*, 18(1): 35–49.

Liben, L.S. (1987) Information processing or Piagetian theory: conflict or congruence? in L.S. Liben (ed.) *Development and Learning: Conflict or Congruence?* Hillsdale, NJ: Lawrence Erlbaum.

McGuinness, C. (1990) Talking about thinking: the role of metacognition in teaching thinking, in K. Gilhooly, M.T.G. Keane, R.H. Logie and G. Erdos (eds) *Lines of Thinking*. New York: John Wiley and Sons.

McLellan, R. (1999) Motivational style, commitment, and cognitive acceleration. Conference paper, Montreal meeting of the American Educational Research Association.

NFER-Nelson (1993) *Non-verbal Reasoning*. Windsor: NFER-Nelson.

*Nickerson, R.S., Perkins, D.N. and Smith, E.E. (1985) *The Teaching of Thinking*. Hillsdale, NJ: Lawrence Erlbaum Associates.

Pascual-Leone, J. (1976) On learning and development, Piagetian style, *Canadian Psychological Review*, 17(4): 270–97.

Pascual-Leone, J. (1984) Attention, dialectic and mental effort: towards an organismic theory of life stages, in M. Commons, F. Richards and C. Armon (eds) *Beyond Formal Operations: Late Adolescent and Adult Cognitive Development*. New York: Praeger.

*Perkins, D. (1995) *Outsmarting IQ*. New York: The Free Press.

Perkins, D.N. and Saloman, G. (1989) Are cognitive skills context bound? *Educational Researcher*, 18(1): 16–25.

Piaget, J. (1950) *The Psychology of Intelligence*. London: Routledge and Kegan Paul.

Piaget, J. and Inhelder, B. (1974) *The Child's Construction of Quantities*. London: Routledge and Kegan Paul.

Raven, J.C. (1960) *Guide to the Standard Progressive Matrices Set A, B, C, D, E*. London: H.K.Lewis.

Sánchez, J.M. (1997) Nature and modes of metacognition, in J.M.M. Beltrán, L.T. Belmonte and J.A.R. Moreno (eds) *Is Intelligence Modifiable?* Madrid: Bruño.

Shayer, M. (1996) *Long-term Effects of Cognitive Acceleration through Science Education on Achievement*. King's College London: Centre for the Advancement of Thinking.

Shayer, M. and Adey, P.S. (1981) *Towards a Science of Science Teaching*. London: Heinemann.

Sternberg, R.J. (1985) *Beyond IQ: A Triarchic Theory of Intelligence*. Cambridge: Cambridge University Press.

*Sternberg, R.J. (1996) Myths, countermyths, and truths about intelligence, *Educational Researcher*, 25(2): 11–16.

Sternberg, R.J. (1998) Abilities are forms of developing expertise, *Educational Researcher*, 27(3): 11–20.

Thorndike, R.L., Hagen, E.P. and Sattler, J.M. (1986) *Stanford-Binet Intelligence Scale*, 4th edn. Riverside: DLM Teaching Resources.

Towse, J., Hitch, G. and Hutton, U. (1998) A re-evaluation of working memory capacity in children, *Journal of Memory and Language*, 39: 195–217.

Voss, J.F. (1989) Problem solving and the educational process, in A. Lesgold and R. Glaser (eds) *Foundations for a Psychology of Education*. Hillsdale, NJ: Lawrence Erlbaum.

*Vygotsky, L.S. (1978) *Mind in Society*. Cambridge, MA: Harvard University Press.

Wechsler, D. (1958) *The Measurement and Appraisal of Adult Intelligence*, 5th edn. Baltimore, MD: Williams and Wilkins.

Woodcock, R.W. (1995) Conceptualizations of intelligence and their implications for education. Conference paper, San Francisco meeting of the American Eductional Research Association.

11 Progression and differentiation

Chris Harrison,
Shirley Simon
and Rod Watson

A science lesson provides learning experiences – experiences which should challenge children's current scientific understanding, and possibly, develop new ideas and questions about the world around them. Ultimately, it is the teacher who decides what these learning experiences should be (Selmes 1974), and how the ideas are conveyed through the activities they provide. Despite the introduction of National Curricula and other external means of specifying the curriculum, decisions regarding the implementation of the curriculum *still reside* with the teacher (Woolnough 1994). Thus, while national or local curricula provide a framework for the curriculum, inform- ing teachers broadly when aspects of a topic should be taught, there is still a great deal that the teachers have to consider in preparing and providing for the learners in their charge.

That there should be continuity and progression within the curriculum is self-evident (Woolnough 1994), but how can this be achieved over the 11 or more years of compulsory schooling? This chapter discusses several examples of proposed sequences which are based on research studies. Their implications will be examined both by specific reference to the English and Welsh National Curriculum, and also as sources for some general guidance on developing ideas of progression and differentiation in the teaching of science.

What do we mean by progression?

There is no single definition of progression. Rather, a range of alternative understandings can be found in the literature. The most common are:

- progression is the order in which a child's learning takes place;
- progression is the sequential difficulty within a topic.

One might assume that the first of these follows from an analysis of the second, in that children will progress in their learning if they are presented with a curriculum which incorporates progressively more complex ideas. However, research shows that children's learning is not straightforward. For, although we may decide on what appears to be a hierarchy of difficulty within a subject, and organize our scheme of work accordingly, children may take different pathways to reach understanding, possibly in unpredictable ways (Denvir and Brown 1986a, 1986b).

In addition to these ideas there are notions of progression arising from psychological studies of cognitive development (Adey 1995, 1997), where:

- progression is the child's ability to process increasingly more complex ideas.

This idea of progression has implications for the previous two, in that the stage of a child's cognitive development will determine the degree of difficulty a child is able to tackle within a topic, and hence, the way in which the child's learning takes place within the topic. Adey (1997) argues that policy-makers have not taken such ideas of progression into account, and suggests that, without a more theoretically justified notion of progression, the implementation of any National Curriculum will continue to be problematic. We return to his arguments later.

Progression has also been defined in terms of conceptual change with respect to specific concepts, for example, in science:

- progression is a shift from a naive conception to a more scientifically acceptable concept.

The idea of progression as conceptual change has underpinned many domain-specific studies of progression, some of which have researched children's ideas at different ages, and others which have studied children's changing conceptions over a period of time.

That progression can be viewed in so many different ways has given rise to much confusion surrounding the term – particularly when planning, teaching, and assessing the outcomes of a statutory curriculum, which is based upon progression and articulated in terms of levels of achievement.

Progression as extent, breadth or conceptual complexity?

Adey (1997) proposes a three-dimensional model to assist in the process of defining progression, whose dimensions are 'conceptual complexity', 'extent' and 'breadth'. The 'breadth' and 'extent' dimensions describe the total amount of knowledge an individual has about subject matter in the curriculum. Adey suggests that there is no absolute way in which progression can be defined by the dimensions of 'extent' and 'breadth'. For, if there is no change in 'conceptual complexity', the order in which pupils learn topics is arbitrary. Progression in the dimensions of 'extent' and 'breadth' can only be measured in terms of the amount of knowledge an individual has, or according to the order in which topics are assessed. On the other hand, the conceptual complexity dimension, articulated from psychological foundations, 'holds the key to a true hierarchical notion of progression'. Adey makes the point that conceptual complexity is more than a concept hierarchy (Gagné 1970), where a concept is analysed in terms of the sub-concepts which are logically prior to it, because such hierarchies fail to analyse concepts in terms of their cognitive demand.

In the mid-1970s, Shayer and his colleagues carried out a large-scale study of students' stages of cognitive development (Shayer *et al.* 1976; Shayer and Wylam 1978). Using Piaget's stage theory of cognitive development, they derived figures for the percentage of children who were capable of thinking in a pre-operational, concrete or formal manner (Table 11.1).

Table 11.1   Percentage of children at different Piagetian stages in a 12,000 case sample representative of British children

Stage	Age 11y 1m (%)	Age 15y 9m (%)
3B (late formal)	0	13
3A (early formal)	6	18
2B (late concrete)	37	54
2A (early concrete)	49	15
1 (pre-operational)	8	0

The early research of Shayer and Adey led to the Cognitive Acceleration in Science project (CASE), from which the *Thinking Science* materials for schools were produced. The *Thinking Science* pack contains 32 activities, each relating to reasoning patterns characteristic of higher-order thinking. An analysis of these activities shows well-defined lines of progression within each reasoning pattern. For example, activities 9, 10 and 11 show clear progression in terms of reasoning, based on compensation, beginning with the simple notion that 'as one variable goes up the other goes down' – found in the relationship between thickness of a tree branch and number of joints from the ground – to the more complex idea of inverse proportionality in the relationship between current and length of wire in a simple electric circuit.

More recently, Adey and Shayer (1993) claim that their research has provided evidence that enhancement of thinking skills has long-term effects that transfer to other subjects outside science, suggesting that a child's cognitive development is independent of subject. Indeed, the *Thinking Science* materials are intended to be used independently of curriculum content, the idea being that the enhancement in children's thinking helps them to access conceptually complex material more readily regardless of the subject. The evidence to support this is that GCSE results of schools taking part in the *Thinking Science* INSET programme are higher than expected (Shayer 1996).

If we accept Adey's compelling arguments, and the evidence accumulated from the CASE research, it could be argued that the notion of progression in terms of 'conceptual complexity' is sufficient, and all we have to do is apply a Piagetian framework to all the learning objectives in the science curriculum in order to arrive at the optimum sequence. Attempts to do this, however, may be confounded by the unpredictable ways in which children come to understand concepts in different parts of science (Chapter 3) and the language of science (Chapter 6). It is therefore useful to contrast and support the ideas and evidence of Shayer and Adey with the results of research which examines the evidence of how children progress in their understanding of specific concepts.

Progression as conceptual change within a domain?

The literature in science education of the 1970s and 80s shows an abundance of studies on children's ideas (see Driver *et al.* 1994) and theories of conceptual change, such as Posner *et al.*'s (1982) theory for the accommodation of a scientific conception. Such accommodation is seen as a radical change in a person's conceptual system, yet something which happens in a gradual or piecemeal fashion:

> Students are unlikely to have at the outset a clear or well-developed grasp of any given theory and what it entails about the world. For them, accommodation may be a process of taking an initial step toward a new conception by accepting some of its claims and then gradually modifying other ideas, as they more fully realize the meaning and implication of these new commitments. Accommodation, particularly for the novice, is best thought of as a gradual adjustment in one's conception, each new adjustment laying the groundwork for further adjustments but where the end result is a substantial reorganisation or change in one's central concepts.
>
> (Posner *et al.* 1982: 223)

The process of conceptual change characterized here has been the focus of many studies concerned with the ways in which students come to understand scientific concepts and reach a scientific view (Pfundt and Duit 1994; Krnel *et al.* 1998). Researchers have attempted to analyse the process of conceptual change to find ways in which teaching and learning science might be supported. For example, Andersson (1986) identified that a common core to pupils' explanations are models of causation developed from everyday experience – elements of which recur in their scientific explanatory models; for instance, the idea that light travels further at night is based on observations that it is easier to see a distant light at night. Clement *et al.* (1989) explore how the identification of *anchoring conceptions*, that is pupils' pre-conceptions which are in agreement with accepted theory, can be used to assist the process of concept development. For example, many pupils do not believe that a table exerts an upward force on a coffee cup sitting on the table, but most pupils believe that a spring will exert a constant force on one's hand as one holds it compressed. In teaching about forces, this intuition about springs can be built on as an anchor. By working with pupils to help them see that even 'rigid' objects are springy to some extent, one can anchor the idea of static forces in the pupil's everyday experiences of springs and forces.

Examples of teaching approaches, which build on research into the process of conceptual change, can be found in a more recent collection of papers (Welford *et al.* 1996). For example, Arnold and Millar's (1996) suggestions for the teaching of heat draw on the use of analogies in teaching science (for example Treagust *et al.* 1992). Arnold and Millar develop a water analogy as part of an introductory course on thermodynamics, where the supply and flow of water models the supply and flow of heat.

The complexity of conceptual change has important messages for those wishing to understand progression within specific domains of science. This complexity is revealed in studies which map children's progression through individual concepts. The large body of research evidence on students' ideas, or *alternative frameworks*, which differ greatly from the scientific view and which are resistant to change by teaching (Driver *et al.* 1994), has implications for notions of continuity and progression when planning teaching experiences to support learning. As Driver *et al.* point out:

> Teaching science with children's thinking in mind depends upon careful planning in which continuity of curriculum is designed for progression in pupils' ideas. The term progression is applied to something that happens inside a learner's head: thinking about experiences and ideas, children develop their ideas. Some aspects of this learning may happen quite quickly and easily, whereas other aspects happen in very small steps, with difficulty and over a number of years. Continuity, on the other hand, is something organised by the teacher: it describes the relationship between experiences, activities and ideas which pupils meet over a period of time, in a curriculum which is structured to support learning. Curricular continuity cannot guarantee progression.
>
> (Driver *et al.* 1994: 12)

Driver *et al.* make the point that the sequencing of the *small steps* can be informed by what is known about progression of children's understanding. In developing INSET materials on progression, the CLISP team (Driver *et al.* 1990) draw on a range of research studies into children's ideas on light and evaporation to show how these ideas change with age, and hence inform our understanding of progression in these areas. A similar attempt to describe progression within a specific area of science was conducted by a survey of 300 students in Spain and England to ascertain the range of explanations these students had about combustion (Prieto *et al.* 1992; Watson *et al.* 1997). Their study led to a model of progression defined by a hierarchy of explanation, ranging from burning and combustion explained by *description* of observations, through notions of it as a process of *modification* and *transmutation*, to the scientific views about chemical reactions.

A limitation of describing progression in this way is that the ideas are snapshots of different children's thinking: they do not arise from longitudinal studies of individual children, like that of Denvir and Brown (1986b), which show how children's ideas progress over time. Unfortunately, longitudinal studies are rare and expensive. One study which included a small longitudinal aspect, coupled with a cross-age analysis of progression, was undertaken for the topic of forces (Simon *et al.* 1994a, 1994b; Black *et al.* 1996). The research approach for this study was founded on the view that learning might be seen as an attempt to bridge two domains, one being the domain of children's understandings of the world around them, the other being the conceptual apparatus of science.

There were two task situations: the first was a heavy box, which could be lifted by pulling up on an elastic band; the second was a bridge, made from card and bricks, which supported a 600 g mass with obvious distortion. These tasks were used in interviews, where children were asked to make predictions, observations and explanations, using their own terms of pushes and pulls, followed by science terms of force and gravity. Forty-eight children were involved in the study, six at each of the ages 6, 8, 10 and 12. Each child was interviewed on two occasions, several months apart, and information about relevant teaching experienced during this interval was also collected.

Taking into account the longitudinal features, the cross-age analysis and the influence of teaching, which was minimal as forces in balance were not then included in the National Curriculum for primary science, the findings showed that a wide range of predictions and explanations, articulated in the children's own terms, were present throughout the age groups in the sample. However, there was an increase in the use of scientific terms with age (Simon *et al.* 1994a).

The results of this research suggested a possible scheme for progression in children's understanding of forces in equilibrium (Black *et al.* 1996). The outline for this was as follows, the points in italics added as a guide to teaching:

First stage: No relevant features predicted or observed; explanations tautologous or naive; no forces identified. *Scientific terms not used.*

Second stage: Focus only on weight in prediction and observation; explanation in terms only of heaviness; weight identified as the only force. *Scientific terms limited to weight and force as weight.*

Third stage: Focus on distortions; distortion forces the only ones iden-
               tified. *Scientific terms relevant to these in conjunction with
               force.*

Fourth stage: Both weight and distortion effects noticed and used to
               predict; both features used in explanations; opposition of
               the two identified in explaining why things don't fall.
               Wide range of scientific terms used appropriately.

Fifth stage: Equality of the opposing effects seen as the distinctive
               feature of equilibrium; existence of distortion forces
               inferred where distortion not visibly evident; ideas used
               over a range of phenomena. *Able to give explanations of the
               scientific terms used.*

(Black *et al.* 1996: 138/9)

A teaching approach was developed based on this outline and tried out in
one school with samples of 6- and 10-year-old children (Simon 1994b). Children of both ages were provided with a range of phenomena, including those
where distortion and perceived force are clearly related (for example, in
compressing a bedspring), those where the link is less obvious (for example,
a weight on a piece of foam rubber) and everyday cases, like a book on a
table, where such features have to be inferred (Clement *et al.* 1989). Children
were encouraged to look at these situations as a whole, focusing on several
interacting features. The use of the term force was expanded from situations
where personal effort was involved (for example, squashing the spring), to
those where observed effects (for example, the distorted foam) were used as
evidence of forces. Pre- and post-interviews showed that most children in
each year group made progress in identifying pushes, pulls and forces as a
result of the teaching approach. Overall, the research showed that focused
teaching based on a well-defined notion of progression, using exploration of
appropriate phenomena to make predictions and explanations, could probe
and enhance young children's thinking about forces in a way that had not
been previously considered. One of the questions raised by this research is:
what is progression like in other concepts in science, and more importantly,
how can curriculum topics best be introduced to children? Some answers
can be found in reviews of the literature such as Driver *et al.* (1994).

Progression of understanding in investigations and the processes of science

Progression of understanding in the procedural aspect of science – how science is done – can be seen from two perspectives: On the one hand, there is
the development of children's capabilities and, on the other hand, there is a
hierarchical progression through the subject.

Progression from a child development perspective

From the child's perspective, the problem is to select learning experiences that take into account the current state of students' development. From the subject perspective, the problem is one of unravelling the internal structure of the subject and organizing it in such a way that students progress through increasingly difficult tasks. The two perspectives are interrelated, but for the purposes of analysing the lessons from the research literature, they will be treated separately here.

We shall start by returning to the work of Shayer and Adey (1981). What are the implications of their findings for developing procedural understanding? A student at level 2A (early concrete) can begin to come to terms with fair tests in simple contexts where effects are intuitively obvious, but may have difficulty in isolating variables and understanding the need to change one independent variable at a time. Students at this stage have difficulty in holding two variables in mind and envisaging relationships between them. For instance, when considering relationships between two variables, they often simplify the problem by simplifying the relations between the variables. For example, when describing the relation between the time taken to dissolve sugar in water of different temperatures, a student may focus on just one pair of values (the hot one dissolves quickly) or extreme pairs of values (the hottest one dissolves most quickly and the coldest one dissolves most slowly) rather than focusing on the overall pattern. 'Fairness' may also be applied in the sense of giving every factor an equal chance, for example a 'slower runner should be given shorter distance' (Shayer and Adey 1981). When the variables being studied are abstract and non-intuitive, such as the relation between particle size of a solid and rate of reaction, it is unlikely that the student will be able to carry out the mental operations needed to set up a fair test. Similarly, in interpreting graphs of relationships between variables, a student at stage 2A may not see the pattern in the graph and, instead, may focus on pairs of results at the extremes of the graph to describe relationships.

A student at stage 3B (late formal operations) has a different quality of thinking, and so now a student:

> Sets up suitable experiments to economically control factors and eliminate ones that are not effective, and can apply 'all other things equal' strategy to multivariate problems. More sophisticated biological experiments possible including interaction effects. Appreciates the impossibility of controlling natural variation, and so the need for proper sampling.
>
> (Shayer and Adey 1981: 79)

This analysis of the effect of child development on performance is supported by performance data. Strang *et al.* (1989) found that performance in investigative tasks was affected by:

- the number of independent variables in a task;
- the nature of the variables in a task, for example whether they are continuous, discrete or categoric;
- the measurements needed to judge the dependent variable: deciding what to measure, how and how often, and, finally, difficulties in making the measurements.

Performance in all three areas increased with age. Conceptual demands were described in terms of pupils' understanding of variables in an investigation, and these were found to have a profound effect on pupils' performance. For example, in an insulation investigation, some pupils thought that the insulating material actually generated heat and so kept the insulated container warm.

They also found that performance in identifying the independent, dependent and control variables, in operationalizing them, and in choosing appropriate values for the independent variable, all increased between ages 12 and 14. Similar results were found by Millar *et al.* (1994).

Using this Piagetian model, Kuhn (1989) proposed a sequence in the development of children's understanding of the relationship between theory and data (See Chapters 4 and 5 of this book). In the early stages of development, children do not consciously separate theory and data. This model was used by Dawson and Rowell (1990) to explore the development of scientific reasoning in Year 11 students. They concluded that very few students operated at the upper end of the developmental spectrum, i.e. demonstrated a clear separation between theory and evidence. The results of Driver *et al.* (1996), using a similar framework, give a more encouraging picture, with only a minority producing a consistent argument relating evidence and explanation at age 9, but rising to more than half at age 16.

Another approach, taken by Millar *et al.* (1994), categorized students' responses or *frames* used by students when working on practical activities. The frames describe students' understanding of the nature and purpose of the investigation task. The four kinds of frame used by students were:

- an engagement frame, in which pupils engaged in activities without obvious plan or purpose;
- a modelling frame, in which they tried to produce a desired effect;
- an engineering frame in which they tried to optimize the effect;
- a scientific frame.

Although the use of the scientific frame increased from ages 9 through 12 to 14, still only a minority were using the scientific frame at age 14.

All these studies show a pattern in the response of students that changes with age. The work of Shayer and Adey (1981) explains why, for less able students, teachers tend to give more support or make tasks more closed (Watson and Wood-Robinson 1998; Watson *et al.* 1999): in short, less able students do not have the cognitive skills to perform an investigation unsupported.

Progression from the subject perspective

Strategies to modify the demands of the investigative tasks to match better the capabilities of students were explored by Fairbrother *et al.* (1992). Six possible investigations for the same topic are given as an example:

A.1 Place one batch of seeds in a dark cupboard, another in the light. Control all the other factors.
Which produces the greenest shoots after a week?
A.2 Find out whether light and temperature affect plant growth.
A.3 Discover the factors affecting plant growth.

B.1 Find out if light affects the production of starch in a leaf of this plant.
B.2 Find out how light affects the process of photosynthesis.
B.3 Investigate the factors affecting the process of photosynthesis.

The openness of the investigations increases going down the two lists. In A.1 the variables are specified and guidance is given on altering the independent variable, in A.2 the variables are specified but no guidance is given on manipulating them, and A.3 is left open. The openness changes the opportunities to develop specific skills and procedures. More decisions are taken by the teacher in investigation A.1 than in A.3, affecting the opportunities for independent planning by pupils. Investigation A.3 may be suitable for pupils with extensive experience and understanding of planning investigations. Investigation A.1 treats light as a categoric variable (light or dark) and so limits the kind of data produced and hence the opportunities for interpretation of the data. Investigation A.3 is more open and so pupils can interpret it in a variety of ways and some may decide to treat light as a continuous variable. Set B is different from set A in that all the investigations include a higher conceptual demand. Jones *et al.* (1992) illustrate how teachers were able to alter the demands of tasks to match the needs of different classes and to differentiate tasks to match the needs of different groups of students within the same class.

Millar *et al.* (1994) also analysed tasks in terms of their procedural and conceptual demands, but defined the procedural demands in terms of manipulative skills, frame (see previous section) and understanding of evidence. The study shows how children's understanding of concepts of evidence changed with age and concludes:

The domain of *scientific* evidence contains ideas which must be taught; it cannot simply be assumed that they will be 'picked up' through experience. First, the purpose of science investigating needs to be made clear to children, through explicit discussion and examples, rather than taken for granted. Then, children's understanding of empirical evidence and of criteria for evaluating the quality of evidence needs to be explicitly

addressed through the curriculum. These are, arguably, the key transferable ideas in this domain – ideas which apply generally across a range
of empirical investigations.

(Millar *et al.* 1994: 245)

There are few studies of strategies used by teachers to help children
develop an understanding of the quality of concepts of evidence. The
ASE–King's Science Investigations in Schools (AKSIS) project is developing
and trialling such strategies, and reports an approach designed to help students construct graphs more accurately, and to use them to examine the
quality of evidence (Goldsworthy *et al.* 1999). One of the key principles in
the approach used is that graphs and data must be linked in a meaningful
way: the graphs should tell a story about the data, which makes sense and
therefore, allow the quality of data to be inspected, permitting interpol-
ations, extrapolations and quantitative predictions. Criteria for quality of
graphs can be made explicit by asking pupils to examine and criticize graphs
which have been drawn incorrectly. Gott and Duggan (1995) suggest activities to help pupils develop the criteria for selecting a suitable choice of range
and interval for the independent variable, and for deciding what kind of
graph to use.

Progression as transfer to different contexts

Progression can also be seen as the ability to apply procedural and conceptual knowledge in increasingly broad contexts. Several studies have shown
that context has an effect on students' performance. Shavelson *et al.* (1992)
found that students' performance varied considerably from task to task. Song
and Black (1991, 1992) found that whether the tasks were set in everyday or
school science had a significant effect, whereas Ruiz-Primo and Shavelson
(1995) found that there were groups of structurally different investigations
that resulted in different performances. Lack of clarity in distinguishing different kinds of investigations has led to problems in teaching (Watson *et al.*
1999). In England and Wales, just one kind of investigation, fair testing, has
come to dominate the science curriculum, to the exclusion of other worth-
while kinds of investigations such as ecological studies, chemical analysis,
classification, technological problems and testing theoretical models. When
other kinds of investigations are included in the curriculum, sometimes procedures of fair-testing are used, even though they may be irrelevant to that
kind of investigation.

All these studies lead to the same conclusion: that the ability to carry out
scientific investigations needs to be developed in a variety of contexts using
a variety of investigations if children's procedural understanding is to progress.

Progression and differentiation in practice

As progression can be seen as a hierarchical progression through the subject and, alternatively, in terms of the developing cognitive capabilities of students, achieving the two approaches within a single classroom can be problematical.

Russell *et al.* (1995) completed an evaluation of the implementation of the Science National Curriculum for KS1 to 3, in which they raised progression and differentiation as two of the main concerns of teachers. The third concern was coverage of the curriculum. It is therefore, not surprising that many science departments, in recent years, have adopted published schemes to provide for their science curriculum, in which the coverage and progression has been worked on by the authors of the scheme. Many such schemes also purport to provide a differentiated approach, but it is difficult to see how this can be achieved when diagnostic testing is not an attribute of such courses and when the differentiation that does exist amounts to little more than similar activities, written with different reading ages in mind. One of the most worrying aspects of implementing a published scheme is that it may cause some teachers to shift the responsibility for differentiation (Harrison 1998) from the teacher's assessment of what goes on in their classroom to the expectations prescribed within textbooks.

Progression and differentiation are intricately linked through the processes of formative assessment (see Chapter 2). While progression sets in place the rungs of the curriculum ladder that the learner has to climb, formative assessment reveals to the teacher which rung her learners are currently on, which they are trying for next, and what help they will require to make their next move. While some students may be on a common rung, it is very unlikely that they will all be. And, their next step will depend not only on the achievement of the current level but also on the necessary experiences to reach other rungs. In some cases, formative assessment may reveal that individual students or groups of students have different problems with a specific concept. (Harrison 1998). For example, some may be confusing the terms current and charge, while others may not be able to recognize whether a circuit set-up shows resistors in series or parallel arrangements. Yet other students may be failing on both or neither of these points. What is the teacher to do? A single learning experience to remediate the learning problems may be difficult, if not impossible, to create. The teacher must incorporate a differentiated approach; her actions must attempt to provide for the needs of each student.

The 1986 DES document, *Better Schools*, lists four markers for an acceptable curriculum as being broad, balanced, relevant and differentiated. Most teachers accept that the students in their charge learn at different rates and in different ways, but there is likely to be far less agreement about how best to respond to this diversity. For differentiation to work in a classroom, the teacher needs to be aware of the needs of individuals so that they can incorporate a planned process of intervention, geared to maximize the potential of each learner (Dickinson and Wright 1993). This is not an easy task. The 1982 Cockcroft Report, on mathematics education, revealed a 'seven-year differ-ence' in the understanding of children entering secondary school, and there is reason to believe that a similar range of capability exists for science. Learn-ers also differ in the way they think, work and perceive the world around them (Head 1979). This will ultimately affect the way that learners respond to, and gain from, the teaching that they receive. Postlethwaite's (1993) book, *Differentiated Science Teaching*, suggests a number of strategies for responding to individual differences, as does Dickinson and Wright's (1993) booklet, *Dif-ferentiation: A Practical Handbook of Classroom Strategies*. Two other books worth perusal are *Differentiation in Action* by Stradling, *et al.* (1991) and *Differentia-tion and the Secondary Curriculum: Debates and Dilemmas* by Hart (1996), which contain a number of case studies and reflections on differentiated practice. Many schools claim that they deal with differentiation by grouping their learners according to pre-achievement, believing that they are reducing the range of abilities within any particular group and therefore, making the task of organizing differentiated work for the class simpler. The most common form of grouping in UK schools is setting, in which students are grouped according to prior attainment in a specific subject. The main reason given by schools for this practice is so that teachers can adapt their pace, style and con-tent to particular ability groups. However, this practice can be problematic, since some teachers regard the set as a group of equal learners rather than as a group with a *narrower range of needs* than a mixed ability class, and there-fore proceed to teach them all in the same style, by the same method and at the same pace (Boaler 1998). Differentiation is, therefore, not planned; teachers tend to teach towards a reference group (Dahloff 1971); and the stu-dents are expected to cope with the teaching. It is only when the reverse action is sought, and teachers fit their teaching to the students' learning, that differentiation can flourish.

Differentiation requires the teacher to direct her teaching according to the diversity of her students and, while an understanding of progression will be important in helping her plan and direct where the learners are heading, it is essential that she utilizes formative practices to construct and drive the learning process. To move in this direction requires more time and effort put into developing such practices in her classroom. However, there is no escap-ing the logic that it is necessary to start teaching where each learner is, to plan goals that each learner can achieve, and to provide the support that enables each learner to succeed.

References

*Adey, P. (1995) Progression in pupils' ideas, in M. Monk and J. Dillon (eds) *Learning to Teach Science*. London: The Falmer Press.
Adey, P. (1997) Dimensions of progression in a curriculum, *The Curriculum Journal*, 8(3): 367–91.

Adey, P. and Shayer, M. (1993) An exploration of long-term far-transfer effects following an extended intervention programme in the high school science curriculum, *Cognition and Instruction*, 11(1): 1–29.

Andersson, B. (1986) The experiential gestalt of causation: a common core to pupils' preconceptions in science, *International Journal of Science Education*, 8(2): 155–71.

Arnold, M. and Millar, R. (1996) Exploring the use of analogy in the teaching of heat, temperature and thermal equilibrium, in G. Welford, J. Osborne and P. Scott (eds) *Research in Science Education in Europe*. London: Falmer Press.

*Black, P., Brown, M., Simon, S. and Blondel, E. (1996) Progression in learning – issues and evidence in mathematics and science, in M. Hughes (ed.) *Teaching and Learning in Changing Times*. Oxford: Blackwell.

*Boaler, J. (1998) Setting, streaming and mixed ability teaching, in J. Dillon and M. Maguire (eds) *Becoming a Teacher*. Buckingham: Open University Press.

Clement, J., Brown, D. and Zietsman, A. (1989) Not all preconceptions are misconceptions: finding 'anchoring conceptions' for grounding instruction on students' intuitions, *International Journal of Science Education*, 11(Special issue): 554–65.

Cockroft, W.H. (1982) *Mathematics Counts: Report of Committee of Inquiry into the Teaching of Mathematics in Schools in England and Wales under the Chairmanship of Dr. W.H. Cockcroft*. London: HMSO.

Dahloff, U. (1971) *Ability Grouping, Content Validity and Curriculum Process Analysis*. New York: Teachers' College Press.

Dawson, C. and Rowell, J. (1990) New data and prior belief: the two facets of scientific reasoning, *Research in Science Education*, 20: 48–56.

Denvir, B. and Brown, M. (1986a) Understanding of number concepts in low attaining 7–9 year olds. Part I: development of descriptive framework and diagnostic instrument, *Educational Studies in Mathematics*, 17: 15–36.

Denvir, B. and Brown, M. (1986b) Understanding of number concepts in low attaining 7–9 year olds. Part II: the teaching studies, *Educational Studies in Mathematics*, 17: 143–64.

*Dickinson, C. and Wright, J. (1993) *Differentiation: A Practical Handbook of Classroom Strategies*. London: NCET.

Driver, R., Brook, A., Hind, D. and Holding, B. (1990) Progression of understanding in science, in K. Johnston (ed.) *CLIS Interactive Teaching in Science: Workshops for Training Courses*. Hatfield: ASE.

Driver, R., Leach, J., Millar, R. and Scott, P. (1996) *Young People's Images of Science*. Buckingham: Open University Press.

Driver, R., Squires, A., Rushworth, P. and Wood-Robinson, V. (1994) *Making Sense of Secondary Science*. London: Routledge.

Fairbrother, R.W., Watson, J.R., Black, P., Jones, A. and Simon, S. (1992) *Open Work in Science: An Inset Pack for Investigations*. Hatfield: Association for Science Education.

Gagné, R.M. (1970) *The Conditions of Learning*. New York: Holt, Rinehart and Winston.

Goldsworthy, A., Watson, J.R and Wood-Robinson, V. (1999) *Getting to Grips with Graphs*. Hatfield: Association for Science Education.

Gott, R. and Duggan, S. (1995) *Investigative Work in the Science Curriculum*. Buckingham: Open University Press.

*Harrison, C.A. (1998) *Differentiation in Theory and Practice*, in J. Dillon and M. Maguire (eds) *Becoming a Teacher*. Buckingham: Open University Press.

Hart, S. (1996) *Differentiation and the Secondary Curriculum: Debates and Dilemmas*. London: Routledge.

Head, J. (1979) Personality and the pursuit of science, *Studies in Science Education*, 6: 23–44.

Jones, A., Simon, S., Black, P.J., Fairbrother, R.W. and Watson, J.R. (1992) *Open Work in Science: Development of Investigations in Schools*. Hatfield: Association for Science Education.

Krnel, D., Watson, R. and Glazar, S.A. (1998) Survey of research related to the development of the concept of 'matter', *International Journal of Science Education*, 20(3): 257–89.

Kuhn, D. (1989) Children and adults as intuitive scientists, *Psychological Review*, 96(4): 674–89.

Millar, R., Lubben, F., Gott, R. and Duggan, S. (1994) Investigating the school science laboratory: conceptual and procedural knowledge and their influence on performance, *Research Papers in Education*, 9(2): 207–48.

Pfundt, H. and Duit, R. (1994) *Bibliography: Students' Alternative Frameworks and Science Education*, 4th edn. Kiel, Germany: Institute for Science Education.

Posner, G., Strike, K., Hewson, P. and Gertzog, W. (1982) Accommodation of a scientific conception: toward a theory of conceptual change, *Science Education*, 66(2): 211–27.

*Postlethwaite, K. (1993) *Differentiated Science Teaching*. Buckingham: Open University Press.

*Prieto, T., Watson, R. and Dillon, J. (1992) Pupils' understanding of combustion, *Research in Science Education*, 22: 331–40.

Rowell, J.A., Dawson, C.J. and Lyndon, H. (1990) Changing misconceptions: a challenge to science educators, *International Journal of Science Education*, 12(2): 167–75.

Ruiz-Primo, M.A. and Shavelson, R. (1995) Rhetoric and reality in science performance assessment: an update. Conference paper, San Francisco Conference of American Educational Research Association.

Russell, T., McGuigan, L. and Qualter, A (1995) Reflections on the implementation of National Curriculum science policy for the 5–14 age range: findings and interpretations from a national evaluation study in England, *International Journal of Science Education*, 17(4): 481–92.

Selmes, C. (1974) Agents for change? in C. Selmes (ed.) *New Movements in the Study of Teaching Biology*. London: Temple Smith.

Shavelson, R.J., Baxter, G.P. and Pine, J. (1992) Performance assessments: political rhetoric and measurement reality, *Educational Researcher*, May: 22–7.

Shayer, M. (1996) Long term effects of Cognitive Acceleration through Science Education on achievement. Unpublished paper, Centre for the Advancement of Thinking, King's College London.

Shayer, M. and Adey, P. (1981) *Towards a Science of Science Teaching*. London: Heinemann Educational Books.

Shayer, M. and Wylam, H. (1978) The distribution of Piagetian stages of thinking in British middle and secondary school children. II – 14- to 16-year olds and sex differentials, *British Journal of Educational Psychology*, 48: 63–70.

Shayer, M., Küchemann, D.E. and Wylam, H. (1976) The distribution of Piagetian stages of thinking in British middle and secondary school children, *British Journal of Educational Psychology*, 46: 164–73.

*Simon, S., Black, P., Brown, M. and Blondel, E. (1994a) Progression in understanding the equilibrium of forces, *Research Papers in Education*, 9(2): 249–80.

Simon, S., Black, P., Blondel, E. and Brown, M. (1994b) *Forces in Balance*. Hatfield: Association for Science Education.

Song, J. and Black, P.J. (1991) The effect of task contexts on pupils' performance in science process skills, *International Journal of Science Education*, 13(1): 49–59.

Song, J. and Black, P.J. (1992) The effect of concept requirements and task contexts on pupils' performance in control of variables, *International Journal of Science Education*, 14(1): 83–93.

*Stradling, R., Saunders, L. and Weston P. (1991) *Differentiation in Action*. London: HMSO.

Strang, J., Daniels, S. and Bell, J. (1989) *Assessment Matters, No. 6: Planning and Carrying out Investigations*. London: SEAC.

Treagust, D., Duit, R., Joslin, P. and Lindauer, I. (1992) Science teachers' use of analogies: observations from classroom practice, *International Journal of Science Education*, 14(4): 413–22.

* Watson, R. and Wood-Robinson, V. (1998) Learning to investigate, in M.Ratcliffe (ed.) *The ASE Guide to Secondary Science Education*. Cheltenham: Stanley Thornes.

Watson, J.R., Goldsworthy, A. and Wood-Robinson, V. (1998) One hundred and twenty hours of practical science investigations: a report of teachers' work with pupils aged 7 to 14. Conference paper, Copenhagen, Royal Danish School of Educational Studies Conference on Practical Work.

Watson, J.R., Goldsworthy, A and Wood-Robinson, V. (1999) What is not fair with investigations, *School Science Review*, 80(292) 101–6.

Watson, J.R., Prieto, T. and Dillon, J. (1997) Consistency in pupils' explanations about combustion, *Science Education*, 81: 425–44.

Welford, G., Osborne, J. and Scott, P. (1996) *Research in Science Education in Europe*. London: Falmer Press.

Woolnough, B.E. (1994) *Effective Science Teaching*. Buckingham: Open University Press.

12 Information and communications technologies: their role and value for science education

Margaret Cox

In 1998 new government programmes were announced which will provide funding of over £1.2 billion towards the National Grid for Learning, ICT for all initial teacher trainees and an ICT training allocation of £460 for every practising teacher. Yet, despite these initiatives and inducements, there is still only a limited uptake of ICT in science education (DfE 1995a; Stephenson 1997), and only a minority of science teachers use ICT regularly in their teaching. Clearly, the majority of science teachers are unconvinced about the value and useful role of ICT.

The power and accessibility of ICT are now narrowing the gap between school science and industrial science, enabling science teaching to provide opportunities for pupils to emulate uses of ICT made by professional scientists. Therefore, consideration needs to be given to the kinds of ICT-based science activities, which will help pupils both to learn science as well as to experience it.

This chapter explains why and how science teaching can benefit from the use of ICT, supported by evidence from many different research studies. Examples are provided of how to use ICT in science lessons, while retaining many of the sound existing pedagogical practices of the good science teacher. Thirty years of research into the effects of IT on the teaching and learning of science has provided substantial evidence of the enhancement of teaching methods and of pupils' learning of science concepts, skills and processes.

Education through science using ICT

Data collected by Woolnough (1994) show that education *through* science can make a positive contribution to the attitude of pupils through enhancing their self-confidence and pride in their work, their autonomy and commitment, depth of thought, presentation and debate; and also, to their communication and general problem-solving skills, cooperation with others and interpersonal skills. Evidence of how these contributions can be further enhanced through the use of ICT has been found through research across a range of subjects. Examples of evidence include:

- taking greater responsibility for their own learning tasks – reported by Crook (1991), who conducted a review of a range of ICT in education research studies including studies in science education;
- organizing their own learning environment – found by Hill (1990), who conducted a case study of pupils using computers in the primary classroom;
- spending longer on the learning tasks – reported by Watson (1993) on the investigation into the impact of IT on pupils' achievements in science, English, mathematics and geography;
- having greater tolerance and empathy – 'the children (travellers and domiciled) all appeared to be much more tolerant of each other and to have begun to appreciate each other's strengths' (Baldwin 1990: 14);
- increasing participation in discussion and using more exploratory language to arrive at choices through discussion (Wild and Braid 1996).

Attitude and motivational changes have been noted with pupils showing:

- increased commitment to the learning tasks for pupils learning science, mathematics, English or geography (Watson 1993);
- enhanced enjoyment and interest in learning of pupils using an artificial intelligent tutor in biology lessons (Schofield *et al.* 1993);
- enhanced sense of achievement in learning among pupils using laptops for a year in science, mathematics, English and geography (Morrison *et al.* 1993);
- increased self-directed learning and independence (Stradling *et al.* 1994) in a survey of 563 primary and secondary school pupils who had used portable computers.

Although the evidence above does not imply that these improvements in attitudes and generic skills will always lead to an improved understanding of science, substantial evidence on pupils' motivation by Ames (1992: 263) and others has shown that 'tasks that involve variety and diversity are more likely to facilitate an interest in learning and mastery orientation.'

In order for teachers to take advantage of the many positive contributions which ICT can make to science education, it is important to understand the many barriers which they also face.

Barriers to the use of ICT in science education

There are some obvious barriers to teachers taking up the use of ICT in science, and some which have been found by research to have a major impact, but which are not so easily recognized.

Today, evidence from school surveys of ICT resources available to science teachers (DfE 1995, and DfEE 1998) and Ofsted reports (Goldstein 1997) shows that a range of school circumstances makes ICT marginal to the science teaching activities. The McKinsey report showed that, in a survey of 550 schools from 1993–94, less than 5 per cent of science teachers were using ICT regularly in their teaching compared with 34 per cent of mathematics teachers (McKinsey 1997). The main arguments for its lack of use, commonly reported by school inspectors (Ofsted) and government commissioned reports, are that:

- the generic software provided with the school's network is all that is needed for using ICT within school subjects;
- the ICT lessons should be independent of other science lessons;
- all pupils must be using the computers all the time during the lesson;
- if the system does not work as anticipated, a whole lesson is wasted;
- ICT is not relevant to the science curriculum;
- ICT does not contribute much to pupils' learning of science.

Even if a science teacher is convinced of the value of ICT to science education, it is difficult to arrange regular use of ICT with a class of pupils. For example, a typical scenario for ICT opportunities in secondary school science teaching is as follows:

> The ICT provision consists of one or more networked rooms, with 15–30 machines running generic software such as the spreadsheet package Excel. The science teacher is expected to plan a lesson around the use of Excel with one pupil per computer, or pupils working in pairs, and because there is a demand for use of the room by other teachers, the science pupils are expected to be using the computers for the entire lesson. This can result in the science teacher's pedagogical practices being influenced primarily by the technology rather than by the knowledge of the best approach for teaching the particular topic.
>
> (Watson 1993: 14)

Solutions: overcoming the barriers

An over-riding problem with using ICT in science education is caused by teachers believing that they have to change their pedagogical practices completely. Research by Harkin *et al.* (in Wubbles 1995) has shown that the behaviour of teachers towards pupils, and the pedagogies they adopt in traditional classes, have a major influence on the cognitive achievements of the pupils. A range of studies reported in the literature, where ICT is also used, reveals changes in some aspects of pedagogy by some target teachers. Examples of these changes are given below.

The computer as a third person

'The computer becomes the third person in a [pupil–teacher] relationship, with a life of its own: with foibles and problems like anyone else which pupils accept' (Edwards 1990: 24). In such circumstances the teacher still has a leading role to play. The problems arise when the teacher *withdraws completely from a leading role in the classroom*, leaving only the pupil–computer relationship.

The teacher as facilitator

Underwood (1988), who reported a study of the pedagogies of teachers using information handling packages in 12 primary schools and six secondary schools, found that many teachers believed that, when using ICT in their lessons, their role would be 'one of classroom manager and facilitator in the learning situation, and they would no longer be the pivotal focus of the classroom'.

Teacher as observer

Evidence of teachers *withdrawing from a leading role in the classroom* when using ICT is supported by the work of many other researchers, including Hoyles *et al.* (1986). They found that, when pupils were given the task of using LOGO in mathematics, the teachers expected to withdraw from guiding the children to select the learning tasks and that there were 'considerable gains in terms of motivation from the pupils deciding upon their own activity'. Similar findings led Chatterton (1988) to conclude that 'the teacher, therefore, should not intervene unnecessarily and should take care that his/her interventions are not needlessly prescriptive'.

Studies of the adoption of ICT into one's teaching have shown that the appropriate degree of pupil autonomy depends upon the nature of the software, the IT expertise of the teacher and pupils, and the relevance of the software and ICT activities to the whole curriculum (Cox and Rhodes 1989; Jackson and Kutnick 1996; Tiberghien and de Vries 1997). On the one hand, there is substantial research evidence which shows that pupils are often given insufficient guidance and leadership from the teacher in order to use ICT effectively (Olson 1992; Reinen 1996). On the other hand, there is substantial research evidence showing that, where the pupils are given some autonomy to investigate and test their own ideas and understanding using ICT, as described earlier, then changes in their learning techniques, and improvements in learning and achievements follow (Papert 1980; Messer and Light 1991; Niedderer *et al.* 1991, Mellar *et al.* 1994).

In practice, it is important for science teachers, when using ICT, to maintain faith in their own teaching expertise when using more traditional science education resources. So, for example, in the case of using a science simulation, this should be related to previous topics and concepts; it should be introduced with clear aims and objectives; and pupils can be given planning tasks and writing-up tasks, as with a laboratory experiment, to help them relate the ICT activity to their other work. This means that an ICT lesson can include time away from the computer, with some of the pupils working with other materials, time for discussion, and with the teacher leading the lesson.

Based on such evidence, ten golden rules to follow when planning and using ICT in science education for its effective use are:

1 *identify* the learning aims and objectives for the pupils which can be enhanced by the use of ICT;
2 *select* appropriate ICT resources to meet the learning aims;
3 *ensure* that the pupils have enough ICT skills to be able to carry out the activity;
4 *plan* the timing of the activity to include non-ICT tasks such as question and answers, group work, pupils' discussions;
5 *plan* enough lessons to enable the activity to be completed;
6 *decide* on the groupings of the pupils – they do not always have to work alone;
7 *introduce* the lesson to all the pupils first before working on any ICT;
8 *intersperse* the ICT activity with whole-class guidance and direction;
9 *allow* enough time for the pupils to reflect and evaluate their achievements at the end of the lesson;
10 *allocate* homework or other assessed work in which the pupils extend their thinking about the activity, and through which you can find out what they have learned.

One of the widely stated fears about using ICT in education is that many pupils know more than their teachers. Even if some pupils, who tend to be in a very small minority, have more expertise in some aspects of ICT, they are unlikely to have the knowledge and expertise in science itself, which the teacher is trying to teach. ICT use in science education should evolve from the *needs of science* and *learning* and not vice versa.

Education in science

Even before the 1980s, educational curriculum development teams were designing and evaluating educational software to enable students to investigate their own ideas about scientific processes through the use of simulations (Cox and Burge 1978; Laurillard 1978; Cox 1984). As a result of this early research into the effects of computer simulations on students' learning, a substantial number of computer simulations and modelling programs were developed for science education and evaluated to find out how these contributed to pupils' learning. Early studies of the impact of computer simulations and modelling on learning by Papert (1980), Kurland and Pea (1983), Cox (1984), Ogborn and Wong (1984), CERI (1987) and many others, found that this type of software enabled pupils to conduct investigations of scientific processes, which were beyond the limits of their mathematical abilities, thereby enabling pupils to construct scientific relationships that more accurately represented the world around them.

The following examples, which are related to specific science subjects, illustrate the teaching and learning attributes of simulations, based upon fixed scientific models embedded in the software.

Biology simulations

One of the limitations to biology investigations in the laboratory is that either many life processes are far too lengthy, such as the effect of parents' dominant and recessive genes on subsequent generations of offspring, or far too complex, such as the effects of nutrition on blood sugar levels in the human body, to be achievable in the school laboratory.

In biology, the interactions between predators and prey form one of the most important features of an ecosystem, and there are now many computer programs which enable pupils to experiment with a simulated predator–prey system. Pupils can study complex interactions using one prey and one predator species. Given the limitations of time for studying such a system, and lack of access to species in their natural habitat, pupils have no other way of investigating these interactions, either in the laboratory or through field work.

Such a simulation model may be restricted to a single species predator and prey interaction, but a more powerful open-ended modelling package such as Model Builder (Booth and Cox 1997) or Excel, described later, would enable the learner to examine other factors, such as the behavioural features which the predators might have, or prey and predator emigration, or immigration to the area under investigation.

Chemistry simulations

In any chemical manufacturing process, a number of decisions have to be made based on economic and environmental considerations as well as on chemical principles. There are many chemical simulations produced for education, including CD-ROMs with photographs and films of different chemical processes. These types of ICT materials, based on complex models, can be used in the chemistry curriculum as laboratory experiments, individual study and group work.

For instance, a simulation of the manufacture of sulphuric acid can show pupils how to achieve a high acid output yet with total sulphur dioxide conversion, preventing the escape of unconverted sulphur dioxide into the atmosphere. As well as investigating the effects of temperature, pressure, and the presence of a catalyst on the manufacturing efficiency, pupils can see the effects of their decisions on pollution of the environment. An important concept for the pupils to understand is the need to balance the chemical efficiency of the process with the costs to the industry and the environmental impact.

Physics simulations

Much of the pioneering work in science simulations for education was first done in physics (see, for example, Bork 1981) to simulate difficult, dangerous or costly scientific experiments. As discussed earlier, for many years, industrial, commercial and research establishments have been using large-scale simulations to plan and experiment with expensive and sometimes dangerous processes. For example, by using a simplified version of a nuclear reactor model for education, the program Nuclear Reactor Simulator (AVP 1991), enables students to study the behaviour of an Advanced Gas Cooled Reactor (AGR), in the relative safety of the classroom!

Evidence from a substantial study of 29 secondary pupils, using five computers over three years, involving the use of simulations in mechanics lessons (O'Shea *et al.* 1993), showed that 'in a relatively short time a significant amount of conceptual change was detected, which showed up on various measures we used. We found that the number of correct responses and explanations based on correct Newtonian theories increased significantly between the pre-test and post-test and delayed post-test.'

Research by Gallop (1995) using a computer-based simulation of rocket trajectories with 12- to 13-year-old pupils to teach them the relationships between force, gravity, mass, acceleration and distance, found that there was a significant improvement in the pupils' understanding of the relationships between these concepts after having used the computer simulation.

Modelling in science education

Simulations, exemplified in the previous section, and modelling are both included in the National Curriculum, but are often confused. Bliss *et al.* (1992) defined the difference between these as: simulations being the exploration of existing models, and modelling involving expression of one's ideas by constructing one's own models.

Research into pupils' use of modelling in science provides the science teacher with some dilemmas. Modelling programs include programming languages such as LOGO, for which there have been widespread claims for many years about its contributions to pupils' learning. Papert (1980), who invented LOGO, found in his research studies during the 1980s, which were later supported by Cathcart's (1990) findings, that pupils using LOGO developed higher-order thinking skills, including problem-solving, hypothesizing and logical reasoning, which could be applied across a range of subjects. However, other researchers, such as Pea *et al.* (1986) found that there was little evidence of significant improvements in pupils' skills and that any improvements were context specific. Niedderer *et al.* (1991) conducted a review of research evidence into the contributions of computer-aided modelling to physics education, by collecting evidence from (a) empirical investigations of students (mis)conceptions, (b) theoretical investigations about understanding and processes of learning, (c) new teaching strategies and case studies in the classroom, and (d) empirical investigations of learning processes based on methods and results of (a) and (b). Their conclusions were that computer-aided modelling at the secondary level:

- does work in normal classroom settings;
- enlarges the set of phenomena for school physics by more complex and realistic examples;
- shifts the focus of instruction to conceptual examinations of physical phenomena;
- supports teaching strategies that put weight on the active involvement of students in the (re-)construction of meaning.

More recently, Linn and Songer (1993), who reviewed the science education literature about cognitive and conceptual change in adolescence, and presented evidence from their case study on the 'Computer as Lab' project (CLP) designed to help middle-school students' knowledge of elementary thermodynamics, concluded that, in particular, 'encouraging students to actively predict the outcomes of experiments and then reconcile these outcomes with their prediction leads to more powerful understanding'.

In the literature, two types of modelling activities are generally distinguished by researchers (Bliss *et al.* 1992): quantitative and qualitative modelling, although a third category: semi-quantitative modelling was included in Bliss *et al.*'s definition. These two types are briefly illustrated below.

Quantitative modelling

One of the first quantitative modelling software packages produced for education was the Dynamic Modelling System (DMS), designed and developed by Ogborn (1985). DMS enabled pupils to put in their own knowledge (model) and then to experiment with it to test both the validity of their model and the consequences of the values chosen. For example, pupils could build a model of the motion of a satellite around the earth from a series of fairly simple relationships, which the computer then calculated by looping through them in an iterative process. By providing initial values for the variables (position coordinates, masses, gravitational constant and initial velocity) and the time increment, pupils could explore the simulations of the motion of the moon around the earth, the motion of the earth or some other planet around the sun and so on by changing these initial values. With the advent of Windows environments, this particular software package is now less commonly used.

The most commonly found quantitative modelling software in schools today is the spreadsheet package, Excel. This is usually 'bundled' in with networks of PC computers sold to schools and is also widely used in commerce and industry. There are many examples of science models, which can be purchased to provide ideas for further modelling (see, for example, Brosnan 1990; Goodfellow 1990), and which help teachers and pupils to develop an understanding of the opportunities which Excel can offer in the science curriculum.

However, one of the difficulties in modelling in Excel for pupils is that they first have to understand the spreadsheet 'metaphor' or representation; that is, what the row and column designations mean, when and where to insert text, numbers, or mathematical equations, and what conditions to include in order to be able to focus on the science model they are building. Research conducted by the Tools for Exploratory Learning Project (Mellar *et al.* 1994) showed that pupils were able to investigate much more complex models if they were provided in simulations rather than if they had to build their own.

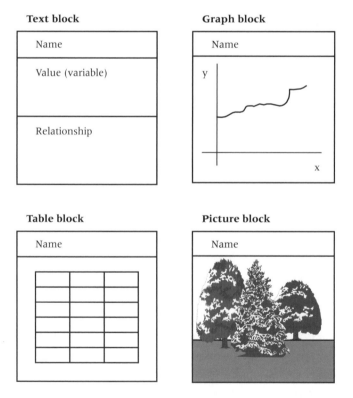

Figure 12.1 Model Builder building blocks

Similar research evidence collected by the Modus project (Webb 1995) and by others such as Seymore Papert (1980), referred to earlier, also showed that the design of the software had a significant impact on pupils' ability to build computer-based models.

One quantitative modelling software package, specifically designed for education to provide a more appropriate interface for the user, is Model Builder (Booth and Cox 1997) which is based on a simple modelling language (incorporating natural language). Figure 12.1 shows the types of blocks that can be created on the screen with Model Builder, which can represent fixed parameters, variables, mathematical relationships, tables, graphs, pictures and video clips (not shown in the Figure). Each of the table, text and graph block's names are incorporated into the modelling language as soon as they are created, enabling pupils to develop complex models sending messages between blocks and constructing sub-models within any text block.

Pupils can create a simple model of a scientific relationship, developing it into an advanced model using sub-models and the iterative process of the software. Schools who have used this software have produced a range of over 80 sample models in many different science topics as well as in other subjects in the school curriculum. These include chemistry models on rates of reaction, physics models on gas laws, radioactivity and energy, and biology models on population growth, predator–prey relationships, and the global carbon cycle.

It is also possible to use the spreadsheet packages Excel or Lotus for quantitative modelling as well as Stella, a semi-quantitative modelling environment.

Qualitative modelling

Although quantitative modelling is the foundation of theoretical science, the advent of the Internet in the 1950s and its phenomenal growth in the last few years (Bagnall 1998) is having a significant impact on society at large. An important aspect of the Internet that is discussed more fully later on in the chapter, which has a common foundation with qualitative modelling, are the decision-making processes involving conditional relationships, such as 'if this is true' and 'if this is true' then 'that is true'. These conditional relationships, based on Boolean logic, form the foundation of search engines, which are used to seek out information on the World Wide Web. Qualitative modelling software environments are usually expert systems or shells based on Boolean logic. The former is a pre-defined set of logical relationships, which enable the user to question the system in order to find out the most likely solution.

A study conducted by Wideman and Owston (1988) of year 7 students, using an expert system shell (a software package which allows the user to design a range of expert systems from scratch) to develop classifications of living organisms, showed that the pupils were able to complete tasks of

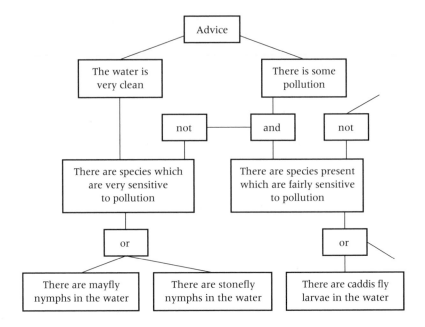

Figure 12.2   Part of an Expert Builder model on water pollution

greater cognitive complexity than is typically required of them at that age, and the externalization of reasoning demanded by the task forced them to employ rigorous and systematic reasoning in order to succeed.

One of the first graphical expert systems shells designed for education was Expert Builder (Webb *et al.* 1993), which enabled pupils to construct logical models using a graphical interface instead of the earlier text-based sequences used in other systems. Figure 12.2 shows part of a model built in Expert Builder by pupils developing a diagnostic system for water pollution using four levels of water quality, measured on its ability to support life. This particular model can be used in various ways, for example, starting from observable features of pollution and considering the possible causes, i.e. starting at the bottom of the diagram model and working upwards. Alternatively, pupils might consider the possible causes of pollution and then evaluate what might be the expected symptoms.

Each condition is typed into a block and is then connected to other conditions through Boolean logical connectors, 'and', 'not', and 'or'. The system has an inference mechanism which enables the learner to query the model at any stage by asking, for example, 'is it true that there are mayfly in the water?'. Learners can query sections of the model or the whole model, block by block.

Teachers and pupils using modelling in science

A recent research study conducted by Webb (1995) of primary school pupils constructing Expert Builder models in science to classify bones, required knowledge 'about the structural arrangement of bones and their functions' and so developed 'propositions about the bones and their positions, and schemas concerning how to identify a particular bone with precision and the knowledge about the function of bones'. Webb found that the pupils learning included:

- extending and consolidating knowledge of the names, characteristics, arrangement and functions of bones and skeletons;
- developing modelling skills;
- developing problem-solving ability through tackling an unfamiliar task;
- developing information retrieval skills involving extracting relevant information from written materials;
- developing the ability to interpret diagrammatic representations and to relate them to concrete structures.

To summarize the discussion above, the main contribution which simulations and modelling can make to pupils' understanding of science is through the acquisition of investigative skills and improved understanding of some scientific concepts and processes. Specifically, research studies have shown that pupils using ICT:

- develop new strategies for solving problems building models by creating entirely new rules (Boohan 1994);
- complete tasks of greater cognitive complexity (Wideman and Owston 1988);
- test personal hypotheses by making predictions (Linn and Songer 1993);
- develop higher-order thinking skills (Cathcart 1990);
- engage in complex causal reasoning (Mellar 1994);
- approach system thinking in a way that is manageable even from a fairly young age (Bliss 1994); and
- use more exploratory language to arrive at choices through discussion (Wild and Braid 1996).

Research into the effects of ICT in science education is not limited to formal years of schooling. There is also evidence of similar positive effects on students in higher education. See, for example, very early work conducted by the Computers in the Undergraduate Science Curriculum Project (McKenzie *et al.* 1978), through to a very recent review of research conducted by Hake (1999). These, and many other similar studies, have shown how higher education students' science learning is enhanced by the use of ICT.

Data handling in science education

ICT activities involving data handling are also gaining great importance because of the availability of large amounts of data on CDs, through the Internet and through the facility provided by sensors for pupils to collect their own data, both locally, remotely and over extended periods of time.

Ever since their invention, computers have had the capability to store and handle large amounts of information, and research has shown that using computer-based data analysis techniques can enhance learners' skills in many aspects of data handling (see, for example, Freeman and Levett 1986; White 1987; Fitzpatrick 1990). These skills include the identification, organization, and retrieval of information. They also demand the high-level thinking skills important for cognitive development of classification (which requires hypothesizing), decision-making, raising questions, and evaluating.

In addition to the cognitive ability of categorization and of retrieving data effectively from a database, Underwood and Underwood (1990) have shown that users need to have some understanding of the nature of data structure. In a study of pupils using database software, Spavold (1989) also showed that understanding the field structure was the most significant factor for successful query formulation. In White's (1987) study with experimental and control groups of secondary school pupils, he showed that the computer-using/ structured-activities group achieved significantly higher mean scores than the non-computer-using/non-structured activities group. He attributed their success to the problem-solving strategies employed for entering, organizing and retrieving data.

In a more recent study, Cox and Nikolopoulou (1997), who investigated the use of data analysis tasks in science lessons, with experimental and control groups of 13- to 15-year-old pupils, to measure their performance in classifying data according to specific criteria, found that pupils using ICT in their data-handling tasks were able to conduct more complex organization and categorization of data than pupils who were using more conventional paper-based techniques.

The development of these skills are very important in science education because these are the generic skills that are valuable in a lot of areas and, as more pupils and teachers gain access to vast amounts of existing scientific data, both off-line through CD-ROMs and on-line over the Internet, using data analysis software will help them develop their capabilities as information seekers and as sorters. For instance, data for foods (carbohydrate, protein, fats, fibre, etc.) can be entered into a spreadsheet from food labels and then sorted to see which foods have the most/least carbohydrate, and so on.

Measurement and data collection

The previous uses of ICT discussed here have involved using stand-alone or networked computers without any peripheral equipment except printers. Yet, one of the most important aspects of science is the use of sensors, switches and other alternative devices, which enable pupils to collect a wide range of scientific data both directly or remotely.

Work by Frost (1994), shows how sensors and switches enabled pupils to become experimental scientists and to compare the data they have collected with their own models of scientific processes. For example, Thornton (1988), studying the effects of using sensors on pupils' understanding of physical phenomena, showed that their knowledge was significantly improved by working with motion sensors. Barton and Rogers (1990), who researched into pupils using motion sensors and light gates to measure the movement of objects, found that, in both cases, 'the immediate presentation of data connects the investigation and the results. This has the effect of freeing pupils to spend most of their time analysing, interpreting and predicting skills, which are at the heart of scientific investigation', rather than wasting large periods of time on the relatively undemanding process of data collection. Practical examples of measurement activities in science include:

- pupils measuring their pulse and changing their rate of physical activity to investigate the relationship between pulse rate and human exertion;
- measuring the rate of reaction between sodium thiosulphate and acid using a light sensor. The experiment can be repeated rapidly to investigate the effects of concentration and temperature;
- measuring the rate of decay of a radioactive substance and using the data analysis package to smooth out random variations in the data.

These and many other similar activities enable pupils to conduct investigations over a long or short period of time and to build systems which enable them to compare their data with their own previous hypotheses of a particular scientific process.

Research discussed earlier has shown how common misconceptions can be overcome through experimentation and computer-based activities. Recently, there are similar studies about pupils' misconceptions concerning topics which overlap with other subjects such as earth sciences and geography. For example, research into pupils' understanding of weather (Dove 1998) has shown that many pupils think that rain is caused when clouds collide; that many 'cannot understand why snow at high altitudes under clear blue skies does not melt' (Dove 1998); and that few children were able to understand high and low pressure. Pupils can investigate many aspects of the weather using sensors measuring one or two factors, or more extensively through the use of an automatic weather station, designed for education. Weather Reporter (Windows on the Weather, 1999) which can be located on the roof of a school, for example, enables pupils to collect hourly information of wind speed, direction, temperature, hours of sunshine, hours of daylight, rainfall, pressure and humidity. By collecting and comparing their data on rainfall, humidity and pressure, for example, pupils can discover the relationship between the three thereby helping them to understand that rain is caused by low air pressure and high humidity levels rather than by colliding clouds.

Conclusions

The different types of ICT in science discussed above are the principal ones of value to science education. There are others, such as Integrated Learning Systems, and drill-and-practice software, but the evidence of the value of these in science education is unconvincing and inconclusive, even though a substantial number of schools are now installing ILS especially for the less able pupils.

There are six main issues concerning the role and value of information and communications technologies in science education.

- First, in science itself, these technologies are becoming a dominant influence on the way in which science is conducted. Therefore, if science education is to be relevant to science in practice, pupils and teachers need experience and confidence in the uses of ICT.
- Secondly, the government is increasing the emphasis on the use of ICT in education, and already, according to government surveys, science teachers are near the bottom end of the 'league' in terms of applying ICT in science education, as compared with other subject teachers.
- Thirdly, there is agreement among many research studies that using ICT can enhance many skills important to becoming a scientist, and which we try and teach our pupils in science.
- Fourthly, there is substantial evidence that appropriate uses of ICT can enhance pupils' interest and motivation and can instil a commitment to learning.
- Fifthly, much of the 'traditional' expertise and pedagogical knowledge of the science teacher is equally relevant when ICT is incorporated into science education.
- Finally, using ICT in science education can make the lessons more exciting and interesting for the teacher as well as for the pupil. Science teachers were probably the first to use experiments in their lessons. Using ICT presents a challenge, but the benefits to teachers, pupils and the future of science education far outweigh the effort required.

References

Ames, C. (1992) Classroom; goals, structures and student motivation, *Journal of Educational Psychology*, 84(3): 261–71.

AVP (1991) *Nuclear Reactor Simulator*. Chepstow: AVP.

Bagnall, P. (1998) The Internet: where it came from, how it works, *Physics Education*, 33(3): 143–8.

Baldwin, C (1990) *Developing Tolerance and Empathy. Developing Pupil Autonomy in Learning with Microcomputers, Teachers Voices*, 4. Coventry: National Council for Educational Technology.

Barton R. and Rogers, L. (1991) The computer as an aid to practical science – studying motion with a computer, *Journal of Computer Assisted Learning*, 7(2): 104–13.

Bliss, J. (1994) Causality and common sense reasoning, in H. Mellar, J. Bliss, R. Boohan, J. Ogborn and C. Tompsett (eds) *Learning with Artificial Worlds: Computer Based Modelling in the Curriculum.* London: The Falmer Press.

Bliss, J., Mellar, H., Ogborn, J. and Nash, C. (1992) *Tools for Exploratory Learning Programme End of Award Report: Technical Report 2. Semi Quantitative Reasoning-Expressive.* London: University of London.

Boohan, R. (1994) Interpreting the world with numbers. An introduction to quantitative modelling, in H. Mellar, J. Bliss, R. Boohan, J. Ogborn and C. Tompsett (eds) *Learning with Artificial Worlds: Computer Based Modelling in the Curriculum.* London: The Falmer Press.

Booth, B. and Cox, M.J. (1997) *Model Builder.* Harpenden: The Modus Project.

Bork, A. (1981) *Learning with Computers.* Bedford, USA: Digital Press.

Brosnan, T (1990) Using spreadsheets in the teaching of chemistry 2: More ideas and some limitations, *School Science Review*, 71(256): 53–9.

Cathcart, W.G. (1990) Effects of Logo instruction on cognitive style, *Journal of Educational Computing Research*, 6(2): 231–42.

*Centre for Educational Research and Innovation (CERI) (1987) *Information Technologies and Basic Learning: Reading, Writing, Science and Mathematics.* Paris: Organisation for Economic Co-operation and Development. France.

Chatterton, J.L. (1988) Knowledge control: the effects of CAL in the Classroom, *Computers and Education*, 12(1): 185–90.

Cox, M.J. (1984) Evaluation and dissemination of science software, *Journal of Science Education in Japan*, 8(2): 147–56.

Cox, M.J. and Burge, E.J. (1978) CAL for Physics, in J. McKenzie, L.R.B. Elton and R. Lewis (eds) *Interactive Computer Graphics in Science Teaching.* Chichester: Ellis Horwood.

*Cox, M.J. and. Nikolopoulou, K. (1997) What information handling skills are promoted by the use of data handling software? *Education and Information Technologies*, 2: 105–20.

Cox, M.J. and Rhodes, V. (1989) *Time for Training – Video.* Swindon: Economic and Social Research Council.

Crook, C. (1991) Computers in the zone of proximal development: implications for evaluation, *Computers and Education*, 17(1): 81–91.

Department for Education (DfE) (1995) *Survey of Information Technology in Schools. Statistical bulletin Issue no. 3/95.* London: The Stationery Office.

Department for Education and Employment (DfEE) (1998) *Survey of Information and Communications Technology in Schools. Statistical bulletin Issue no. 11/98.* London: The Stationery Office.

Dove, J. (1998) Alternative conceptions about the weather, *School Science Review*, 79(289): 65–9.

Edwards, A. (1990) *The Computer: A Modern Aid to Reading. Developing Pupil Autonomy in Learning with Microcomputers. Teachers Voices, 3.* Coventry: National Council for Educational Technology.

Fitzpatrick, C. (1990) Computers in geography instruction, *Journal of Geography*, 89(4): 148–9.

Freeman, D. and Levett, J. (1986) QUEST – two curriculum projects: perspectives, practice and evidence, *Computer Education*, 10(1): 55–9.

*Frost, R. (1994) *The IT in Secondary Science Book: A Compendium of Ideas for Using Information Technology in Science.* London: IT in Science.

Gallop, R. (1995) An investigation into the use of the repertory grid technique to identify links in pupils' conceptual frameworks and changes brought about by the use of a computer simulation. Unpublished PhD thesis, University of London.

*Goldstein, G. (1997) *Information Technology in English Schools: A Commentary on Inspection Findings 1995–1996*. London: The Stationery Office.

Goodfellow, T. (1990) Powerful tools in science education, *School Science Review*, 71(257): 47–57.

Hake, R.R. (1999) American Association for Higher Education. http://www.aahe.org/hake.htm

Hill, J. (1990) *Children in Control: Attempts at Fostering an Autonomous Attitude towards the Computer Work in the Classroom. Developing Pupil Autonomy in Learning with Microcomputers. Teachers Voices, 2*. Coventry: National Council for Educational Technology.

Hoyles, C., Sutherland, R. and Evans, J. (1986) Using LOGO in the mathematics classroom. What are the implications of pupil devised goals? *Computers and Education*, 1: 61–72.

Jackson, A. and Kutnick, P. (1996) Groupwork and computers: task type and children's performance, *Journal of Computer Assisted Learning*, 12(3): 162–171.

Kurland, D.M. and Pea, R.D. (1983) Children's mental models of recursive LOGO programs: *Proceedings of the Fifth Annual Cognitive Science Society*. Rochester, New York: Lawrence Erlbaum Associates.

Laurillard, D.M. (1978) Evaluation of student learning in CAL, *Computers and Education*, 2: 259–63.

Linn, M.C. and Songer, N.B. (1993) 'Cognitive and conceptual change in adolescence' in D. Edwards, E. Scanlon and D. West (eds) *Teaching, Learning and Assessment in Science Education*. Buckingham: Open University Press.

McKenzie, J., Elton, L.R.B. and Lewis, R. (eds) (1978) *Interactive Computer Graphics in Science Teaching*. Chichester: Ellis Horwood.

McKinsey and Company (1997) *The Future of Information Technology in Schools*. London: McKinsey and Company.

*Mellar, H. (1994) Towards a modelling curriculum, in H. Mellar, J. Bliss, R. Boohan, J. Ogborn and C. Tompsett (eds) *Learning with Artificial Worlds: Computer Based Modelling in the Curriculum*. London: The Falmer Press.

Mellar, H., Bliss, J., Boohan, R., Ogborn, J. and Tompsett, C. (1994) *Learning with Artificial Worlds: Computer Based Modelling in the Curriculum*. London: The Falmer Press.

Messer, D. and Light, P. (1991) The role of collaboration and feedback in children's computer based learning, *Journal of Computer Assisted Learning*, 7(2): 156–9.

Morrison, H., Gardner, J., Reilly, C. and McNally, H. (1993) The impact of portable computers on pupils' attitude to study, *Journal for Computer Assisted Learning*, 9(3): 130–41.

Niedderer, H., Schecker, H. and Bethge, T. (1991) The role of computer-aided modelling in learning physics, *Journal of Computer Assisted Learning*, 7: 84–95.

Ogborn, J. (1985) *The Dynamic Modelling System*. Chepstow: AVP.

Ogborn, J. and Wong, D. (1984) A microcomputer dynamic modelling system, *Physics Education*, 10: 138–42.

Olson, J. (1992) Trojan horse: or teacher's pet? Computers and the teacher's influence, *International Journal of Educational Research*, 17(1): 77–84.

O'Shea, T., Scanlon, E., Byard, M. *et al.* (1993) Twenty-nine children, five computers and a teacher, in D. Edwards, E. Scanlon and D. West (eds) *Teaching, Learning and Assessment in Science Education*. Buckingham: Open University Press.

* Papert, S.A. (1980) *Mindstorms: Children, Computers and Powerful Ideas*. New York: Basic Books.

Pea, R.D., Kurland, M. and Hawkins, J. (1986) LOGO and the development of thinking skills, in K. Sheingold (ed.) *Mirrors of Mind*. Norwood, NJ: Ablex.

Reinen, I.J. (1996) Teachers and computer use: the process of integrating IT in the curriculum. Unpublished thesis, University of Twente, Enschede.

Schofield, J.W., Eurich-Fulcer, R. and Britt, C.L. (1993) Teachers, computer tutors and teaching: the artificial intelligent tutor as an agent for change, *Journal of Educational Technology*, 31(3): 579–607.

Spavold, J. (1989) Children and databases: an analysis of data entry and query formulation, *Journal for Computer Assisted Learning*, 5(3): 145–60.

Stephenson, D. (1997) *Information and Communications Technology in UK Schools. An Independent Inquiry*. London: Independent Inquiry into the Use of IT in Schools.

Stradling, B., Sims, D. and Jamison, J. (1994) *Portable Computers Pilot Evaluation Report*. Coventry: National Council for Educational Technology.

Thornton, R.K. (1988) Tools for scientific thinking: learning physical concepts with real-time laboratory measurement tools, in J. Risley and E.F. Reddish (eds) *Computers in Physics Instruction: Proceedings*. Redwood, CA: Addison Wesley.

*Tiberghien, A. and de Vries, E. (1997) Relating characteristics of teaching situations to learner activities, *Journal of Computer Assisted Learning*, 13: 163–74.

Underwood, J.D.M. (1988) An investigation of teacher intents and classroom outcomes in the use of information-handling packages, *Computers and Education*, 12(1): 91–100.

*Underwood, J.D.M. and Underwood, G. (1990) *Computers and Learning: Helping Children Acquire Thinking Skills*. Oxford: Blackwell.

*Watson, D.M. (ed.) (1993) *The ImpacT Report – An Evaluation of the Impact of Information Technology on Children's Achievements in Primary and Secondary Schools*. London: King's College London.

Weather Report (1999) *Weather Report Update*. Hatfield: The Advisory Unit.

Webb, M. (1995) Children building models. Unpublished PhD thesis, Open University, Milton Keynes.

Webb, M., Booth, B., Cox, M.J and Robbins, P.A.J. (1993) *Expert Builder*. Harpenden: The Modus Project.

White, C. (1987) Developing information processing skills through structured activities with a computerised file-management program, *Journal of Educational Computing Research*, 3(1): 355–75.

Wideman, H.H. and Owston, R.D. (1988) Student development of an expert system: a case study, *Journal of Computer Based Instruction*, 15(3): 88–94.

*Wild, M. and Braid, P. (1996) Children's talk in co-operative groups, *Journal for Computer Assisted Learning*, 12(4): 216–31.

Windows on the Weather (1999) *Weather Report Information*. Hatfield: The Advisory Unit.

Woolnough, B.E. (1994) *Effective Science Teaching*. Buckingham: Open University Press.

Wubbles, T. (1995) An interpersonal perspective on teacher behaviour in the classroom. Conference paper, Bath, meeting of the European Conference on Educational Research.

Part III
The science world

The last part of our book contains the fewest chapters: just two. And therein lies a paradox. For science education has to have some general aims, other than its self-perpetuation, and yet science teachers find it quite difficult to think beyond the confines of the classroom and the laboratory. Science teachers and scientists often claim that science is a great cultural achievement, the mainstay of our economy, personally fascinating. But, still the schemes of work that are devised corporately by science departments and which fill science prep-room filing cabinets up and down the country, are very inward looking, traditional and often lacking in anything other than the self-perpetuation of the science teacher's textbook knowledge.

Snatched conversations with Year 10 students after work experience can be the most that science teachers glimpse of a world of work, a world of science, that is beyond the school gates. Chapter 13, written by Joan Solomon and Sally Johnson, reports on the outcomes and effects of a relatively novel vocational course with some additional novel data. Joan and Sally have researched the General National Vocational Qualification (GNVQ) in science and report their findings here for the first time. The information brought back from teachers and students doing such courses transcends GNVQ itself and informs us on the learning of science in general. What makes GNVQ attractive to teachers and students alike provides vital clues as to what makes good practice in science teaching generally.

What Joan Solomon and Sally Johnson's chapter does with respect to place – students being prepared for the world of work outside the confines of the school – Jonathan Osborne's chapter does with respect to time. For, in Chapter 14, Jonathan looks at what sort of science education one might want to consider as being suitable for future citizens. This question is a good place from which to start a book on 'good practice in science teaching' because, until one has thought carefully about one's aims for science education, all the planning of activities and learning can be in vain. So Chapter 14 closes the circle and brings us back to reconsider why we are doing what we do, and in so doing, the importance of research evidence to that endeavour.

13 GNVQ Science at Advanced level: a new kind of course

Joan Solomon and Sally Johnson

The fuzzy overlap between science and technology in the workplace has occasionally been paralleled by attempts to design and run vocational courses with a science content. Such programmes of study offer interesting insights into how science education can be quite different from the usual academic fare of GCSE and GCE A level. This chapter looks at the impact of the General National Vocational Qualification (GNVQ England, Northern Ireland and Wales) in Science at Advanced level some five years after it was first introduced. It does so to illustrate both the value and difficulties of such an approach to science education. The intention to produce a new course was announced in the government White Paper *Education and Training in the 21st Century* in May 1991, only two years before it started in schools and Further Education (FE) colleges. Bearing in mind that its uptake continues to be small – about 2500 students in 1998 as compared with some 25,000 students studying science at A level – one might be tempted to attribute such a small uptake to the short lead time for what is clearly an interesting and quite new kind of course. However, other GNVQs at this level, such as those in business studies, health and social care, art and design, and leisure and tourism have done much better in terms of numbers enrolling. Perhaps low recruitment may, in part, be associated with the overall unpopularity of science at this present time (Smithers and Robinson 1995). But, this is by no means the whole story.

A lack of preparation lead time has not been the main problem. On the contrary, it seems that the government policy advisers courted disaster from the very start by running together some four main purposes to be served by the new qualification, which made it hard for it to succeed. The purposes were mostly high-minded in themselves, in some cases novel and valuable, and in others only constrained by external factors, but it is not difficult to appreciate that they could not all succeed. The course remains on the books because it appeals, as we shall see, to some tutors and students for reasons which are certainly valid for the students' purposes, but hard to realize in practice.

This chapter will begin by looking at these four main purposes, which it was hoped that GNVQ would achieve. It will use evidence, reported here for the first time, from tutors and students to illustrate this and describe how taking an advanced GNVQ in science appears to those involved. Finally, it will shift the emphasis towards management, and will show what can be achieved when GNVQ is being introduced, despite all the inherent difficulties of this rather unwisely designed course.

Vocational courses and 'parity of esteem'

In what sort of esteem do we, in Britain, hold vocational courses? It is worth pointing out that the difference we make between even the words *vocational* and *professional* education is mirrored in few other European countries. In French and German the same word is used to describe both of them. Reflection suggests that it is only a kind of snobbism, a lack of parity in esteem between being a doctor or a mechanic, which underlies this linguistic difference. Of course, it is true that the Ecole des Mines and Ecole Polytechnique in France have very high prestige, and yet neither miners or car mechanics exit from their abstract and intellectual courses. We appear to value all kinds of practical work less, keep more abstract courses for those taking physics, and the more practical ones for engineers, although we know well that understanding all forms of technology now requires a substantial theoretical basis.

The English inventors of the steam engine, which put us in the forefront of the Industrial Revolution, were poorly educated, if at all. We know that the Cornishman, Trevithic, could not write at all, since his patent has no more than a cross against his name. Newcomen was not much better. When the French mathematician Carnot began the first-ever study of thermodynamics a century later, he dedicated his important book (the *Motive Power of Fire*) to these English engineers. One wonders if he knew what illiterate men they had been!

The last two paragraphs show just how unlikely it would be that vocational education and academic education could be accorded the same esteem in our country. The 1944 Education Act had started by putting three kinds of secondary school into operation – the grammar, the vocational, and the secondary modern. We know that it was the second of these which first fell out of use, leaving the secondary moderns to mop up the technically gifted, along with all the others who had failed the 11+ examination. Those who teach in our modern (FE) technical colleges often try to promote a higher image for the science they teach but, like schools and sixth-form colleges, their less able students, as judged by GCSE results, usually get advised to take the vocational GNVQ, which does the status of that course no favours.

Such deeply embedded cultural attitudes are almost impossible to shift, despite the strongest ministerial pronouncements. When John Patten launched the GNVQ, he spoke of it simply as *an alternative route to higher education*, adding that he looked forward to the time when it would become the main qualification for higher education, replacing A level to a considerable extent. That was in 1991. Since then the old arguments about A level being the *gold standard* by which the excellence of all other courses should be measured, and the reduction of targets for university entry from 50 per cent (backed by the Confederation of British Industry) to 30 per cent, effectively turned the clock back. A level was the best, and GNVQ was a vocational alternative. Some have held fast to the notion that at least part of the 18+ qualification choice (currently a smorgesbröd of A level, AS level, GNVQ and even NVQ) should be vocational in order to increase the employability of students. In science this reasoning, as we shall see, is also suspect.

Is it easier?

If the previous section sounds too damning, it is because one of the objectives which teachers and parents find particularly appealing has not been mentioned at all. Suppose GNVQ really is an *easier* alternative to A level (which can be argued to have the most forbidding standard of any school-leaving science qualification in Europe), that would have great advantages for any students who only started learning seriously towards the end of their compulsory schooling. Maybe it would prevent such youngsters from dropping out altogether from education, and even help some of them on to the ladder which adult Access students struggle to grasp later on. Here is evidence from a remarkably honest young woman who was studying GNVQ science in the pilot year (1993–94): 'Really I was terrible at school. All my results were appalling in science . . . I wasn't a good student until the fourth year (Y 10) when I started being serious. I wasn't into science until the fifth year, when I quite liked science' (Solomon 1994: 9).

That student passed her GNVQ (advanced) in science, became the class representative at FE college, and gained entrance to university from which she has just graduated.

Apart from factors like late maturation, which apply to only a few of our students, being *easier* may mean at least four different things.

- First of all, it could just mean that the *subject matter is less, or of a lower standard*. However, it was never intended that academic and vocational courses would be comparable in this simple curriculum way. Vocational courses should include material relevant to performance at work, while academic courses would pursue other, and possibly more abstract, goals.
- Secondly, it could mean that the course would be taught in a way which the students find *easier to understand*. This is an aspect of GNVQ, which will be explored in a later section, under the heading of learning and skills.

- Thirdly, *easier* might mean *easier to gain access* to the course, implying that GNVQ was tailored to students who had achieved poorly at GCSE and this had, possibly quite unfairly, made their schools unwilling to take them on for A level.
- Fourthly, it is possible that *easier* might turn out, in the students' eyes, to mean *easier to pass because the assessment methods were more congenial*.

The third of these alternatives is certainly true. Almost all schools and colleges which run GNVQ courses in science at advanced level also run A level courses. They advise and select students for each of these on the basis of their GCSE results, even though students often beg to be allowed to take A level on the grounds that, as they believe, employers and university admissions officers give preference to A level qualifications (Hayward and Solomon 1993). The tutors usually hold firm, often setting the dividing line at grade C and above, in all sciences taken at GCSE. Are they right in this, or is it no more than a kind of scholarly elitism, once more reflecting the lack of parity of esteem?

Thanks to the ALIS project at Newcastle University, we now have data suggesting that, in a general way, the tutors are right. The chances of getting a

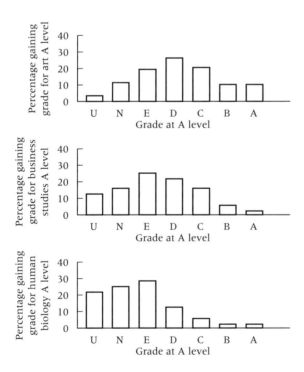

Figure 13.1   Success at A level for those gaining a grade D at GCSE

Source: Fitz-Gibbon (1997: 53)

good grade at A level after achieving a D grade at GCSE are small. As Fig. 3.1 shows, they are less for one of the 'easiest' science subjects than for business studies or art.

We conclude from this evidence that the tutors may indeed be justified in dissuading weaker students from trying to take A level. The same author has explored the connection between UCAS points for university entrance for matched samples (having the same GCSE grades) of GNVQ and A level students. She concludes that the 'equivalence' suggested by National Council for Vocational Qualifications (NCVQ) that a GNVQ grade of 'Distinction' was equivalent to an A and a B, or two As at A level, was not justified by the data. Not only is it easier to gain access to GNVQ courses than to A level ones at school or college, it is also likely to be easier to gain Universities and Colleges Admissions Services (UCAS) points for university entrance from GNVQ.

Performance, outcomes and assessment criteria

We are left with the fourth way in which a course might be preferred – seem easier – which concerns the method of assessment. Gilbert Jessup (1991) proposed a way of setting out the learning objectives of a training course which would, he claimed, make them more transparent to the students and easier to assess. First, he argued that a vocational qualification should be openly concerned with *competence* in a particular vocational field, secondly, that the successful student's *performance* should be judged against clearly defined *criteria*, and thirdly, that there would be both *knowledge and skills* underpinning these performances. To effect this, Jessup, at that time Director of the NCVQ, wanted not only division into units and elements, but also further sub-division into performance criteria which could all be assessed separately.

We can see some of these dimensions on the pages of the GNVQ specifications. There are *units with vocational headings* (sub-divided into elements), and, within these, there are *performance criteria*. If a student (or an employee) is being trained in the use of new equipment the *outcome* of the training will be a *performance* which can easily be assessed if the student is *producing* a product, *identifying* regulations, or *recording* the data collected. All of this makes the students' tasks more straightforward, and their assessment dissectible into separate items in separate units. There is little doubt from interview data collected in 1998 that this is the aspect of GNVQ which still appeals most strongly to a great number of the students, especially when compared with the assessment of A levels. The students were simply asked 'Why did you take this particular course?'.

> There are not any exams – I am terrible at exams – so it's a better course for me. You know where you stand after each coursework.

> Last year I did A level (maths, biology and physics) and failed the three end-of-year exams. So I got kicked off the courses. I heard about GNVQ

and checked it out and thought 'Good, I can stay at the college and do a course I will be good at'. I am finding the work quite easy because there is no stress of exams. I can get on with the coursework and I'm really good at that so far.

Another and rather incongruous performance criterion is *explaining*. Here, the tutors know that they have to go back to teaching knowledge and ensuring understanding. They often said that this was quite welcome and that, 'it gave them space' to be teachers. The examination boards had issued guidebooks which were quite clear and precise about the depth of understanding required. Here again, the knowledge and explanation were in small chunks which the students preferred: 'It is less complicated than GCSE – the way they tell it is easier to understand. There were a lot of things I did not understand in GCSE – this year I do.'

This comparison with the 16+ examination is striking. How can GNVQ at Advanced level hold parity of esteem with A level if it is found to be easier than even GCSE? Reflection suggests that learning in small chunks, rather like the operant conditioning which B.F. Skinner advocated, is bound to be easier at a surface level of learning. But at a deeper level, where understanding is equated to fitting knowledge together (Biggs 1976; Burns *et al.* 1991), it could be argued that small chunks of knowledge cannot be added up to a complete conceptual understanding.

In science, we typically have abstract overarching theories, such as the conservation of energy and mass, the energetics of chemical reactions, and the structure and recombination of DNA. What is taught in GNVQ is likely to be only its application in one or two particular contexts. Will this be enough for studying science at university? And, if it is not, how could a practical scientist or technologist operate in the changing circumstances of modern employment without a broad understanding applicable to new situations?

Certainly Jessup's *New Model of Education and Training*, couched in terms of these discontinuous outcomes, has provoked quite heated controversy in Higher Education. Ron Barnett (1992) is foremost in opposition. Like many others, he is concerned 'that its actual exposition in this case, being based on industry-led standards of performance amounts to an attempt to import into one human activity (education) the interests and conceptions of another (industry)' (Barnett 1992: 7).

Barnett does not deny the obligations of Higher Education to society, in general, but he concludes from this that the quality of student learning is also a matter of public interest. He quotes from the Council for National Academic Awards (CNAA), the body which controlled curriculum innovation in the old polytechnics that 'each programme of study should stimulate . . . an enquiring, analytical and creative approach, encourage independent judgement and critical self-awareness' (Barnett 1992: 13).

GNVQs could, perhaps, do this too to some extent. However, one cannot avoid the judgement that, since they have been devised, and are used, to provide the easier kind of learning, which is bound to the context of its vocational units and measured by specified performance outcomes, this is unlikely. On the other hand, there is very little evidence that these discrete pieces of knowledge prove unsatisfying to the learning student. Both tutors and students saw its advantages.

> The pressure isn't there (in GNVQ) so, although you have a scheme of work, you can make detours if there is a failure to understand something properly, in a way that you don't have time for in A level.
>
> Tutor

> You are doing it stage by stage . . . you can retake the tests (if you don't understand).
>
> Student

Most of the students who were interviewed had been encouraged to take GNVQ because of the poor passes they had gained at GCSE. Just occasionally in the pilot year, students with high grades found themselves taking GNVQ because of administrative problems. Some of these complained that there was too much practical work and too little theory – a plus for the average GNVQ student! – and one added that it was all *too bitty* in conceptual terms. What is encouragingly easy to understand for one student, may have too little conceptual depth for another. The two courses may both be excellent, but they are aimed at students with different strengths, purposes, and cognitive abilities. We do not yet know how far the GNVQ students will catch up with those who have surmounted the challenges of A level, but we cannot really be surprised that most of the more academic universities demand at least one A level as well as GNVQ in their applicants. This may not be *parity of esteem* but, as we shall see, there are places for successful GNVQ students in the newer universities.

Work and vocationalism

The most curious feature of GNVQ is its disconnection from employment. There are two aspects to this. On the one hand, employers take no direct part in the course delivery, on the other hand, there is no compulsion to include work experience, even for a short time, during the two-year GNVQ course. Of course, the second factor is probably imposed by the difficulty in finding opportunities for appropriate work experience in both inner city and rural locations. However, the fact remains that it is odd, to put it mildly, to find no compulsory vocational activity in a vocational course.

Before GNVQ was introduced, there were a number of ways in which students on a vocational course could obtain real work experience:

- apprenticeship on the job – including day-release courses at an FE college with or without craft qualification;

- part-time study for Business and Technician Education Council (BTEC) Certificate, Diploma, usually one day per week at FE college;
- part-time study for Higher National Certificate (HNC), usually one day per week at FE college or Polytechnic;
- a sandwich year in a degree course, which was CNAA validated.

The funding of universities, the policies of the trade unions, the mobility of workers, and hence the reluctance of employers to invest in training, the fall of employment in manufacturing industries, and the rise of employment in the service sector, have all contributed to the decline in popularity of such options. Apprenticeships have all but died out; CNAA ceased to exist when the polytechnics became the new universities; part-time BTEC was phased out in favour of full-time GNVQ, and HNC is now very rare indeed.

Some writers, such as Prais (1995) have argued that we in Britain should look to the *dual support* system of training carried out in Germany as a model for producing more workers with vocational qualifications in Britain. He uses a variety of statistics in an attempt to show an intimate connection between manufacturing productivity and training or education. However it seems equally likely, in the light of the recent change in economic climate across the world, that this is not necessarily a relevant connection, or a predictor of economic success. The present recession in Japan, where about a quarter of all 15- to 18-year-olds attend full-time vocational school, is a sad indictment of the failure of such an approach to provide full employment – a system on which Prais's argument placed so much reliance. Even in Germany, current economic realities are such that employers are showing resistance to paying for out-of-company training, something they had previously done. Rather, as discussed in the opening section, the emphasis on vocational education is more likely to underscore cultural and even linguistic variation than to illuminate any determinants for economic health and productivity.

The reason that the demise of part-time courses in BTEC, HNC and sandwich degree courses, which included work experience, produced regrets is therefore, not economic nor concern at the paucity of vocationally qualified students. Rather, it was that such courses produced a very positive attitude towards study in students. In the course of visiting a large number of new universities, and FE colleges, in order to assess the likely future for GNVQ (Solomon 1994, 1996), several lecturers attested to the enthusiasm with which part-time students tackled their work. Sometimes, they crammed lectures and practical sessions in from 9.00 a.m. to 9.00 p.m. on their one study day. These students, some of them older than the typical post-16 college student, reacted with obvious irritation whenever a lecture had to be cancelled. The contrast with full-time students was, their tutors said, quite remarkable!

Why should the work background have such an effect on learning? The answer may be understood theoretically and detailed empirically. Recent work by Lave and Wenger (1991) has shown that being a part of a working community, however peripherally, produces a change of self-image as well as motivation. Of course, this is as true of college as it is of an industrial science laboratory. It follows from this that no amount of 'doing science' as a student can simulate the skills, and social identity of *doing science* as a worker. Lave and Wenger wrote that college or school 'is a site of specialised everyday activity – not a site where universal knowledge is transmitted . . . Learning is not an epistemological process but a stage in becoming a social person, with skills subsumed in the process' (Lave and Wenger 1991: 40).

In the previous section the question of learning science in or out of context arose because of the greater ease of learning when the knowledge is applied to a particular context. When we face a working and learning situation, it seems that it is not only cognition, but also the values that build up self-identity, which are affected.

This is not a completely new approach to knowledge. In 1982 the CNAA wrote a report for the use of all course teams offering degree courses with supervised work experience (sandwich courses). Their first conclusion ran as follows: 'the primary concern in placing students outside college is their socialisation into occupational and professional values: the enhancing of technical skills and knowledge is of minor concern' (Callaway 1982: ii).

Although not compulsory, Fitz-Gibbon (1997) has shown that, on the whole, GNVQ students were more likely to undertake work experience than were A level students, and that the numbers varied widely from subject to subject, with GNVQ in *Health and Social Care* recording the highest numbers of students taking on work experience during the course. Data were not available for GNVQ science students. However, there was some evidence among the tutors interviewed for this chapter that work experience was very limited indeed in science. One tutor reported that, when the member of staff at the college who was responsible for work experience was made redundant due to funding cuts, the whole effort to find work placements was abandoned. Nevertheless, some GNVQ science students found work for themselves. This lack of readily available work reflects an even more serious lack of permanent employment chances in science for those with Advanced GNVQ, A level, or even first degrees (Solomon 1996). The result of this is that science at GNVQ has, indeed, in the words of John Patten, become an alternative route into Higher Education *and only that*.

> There are scientific jobs (for work experience) in this area in hospitals and companies like SmithKline Beecham, but they would need a degree to get in permanently. So GNVQ Advanced is not an entry point for the jobs they go to on work experience . . . GNVQ is an alternative to doing A level which suits some people better.
>
> Tutor

New skills

Education has long been condemned to a search for the Holy Grail of *transferability*. This becomes even more serious for vocational education in a world where workers may have to be re-trained several times in the course of a working lifetime. And, if vocational education, of the type provided by GNVQ, seems to have become rather disconnected from the actual realm of work as we have seen, then the approach to education which emphasizes generic skills becomes even more important. A good many educationalists are highly sceptical of the whole idea of transferability. Hyland and Johnson (1998) argue that, on both logical and empirical grounds, the notion of domain-independent skills is entirely mythical. Certainly transferability has yet to be demonstrated, although intellectual doubts of this kind seldom influence thoroughly entrenched opinion!

The Manpower Services Commission's New Training Initiative (NTI) began the slippery use of new vocabulary to describe vocational education in 1981, *skills, competencies, inputs, outcomes, evidence indicators* and *performance criteria*, which made its official pronouncements almost incomprehensible at that time. In 1989, the Confederation of British Industry proposed the development of *Core Skills* which might make all education more relevant to employment. The government followed this lead by asking the National Curriculum Council and the National Council for Vocational Qualifications to work out a list of such skills, which they did in 1990:

- numeracy (later application of number);
- modern language competence (later just communication);
- information technology.

None of these three skills is novel to us now, and each is transferable in the sense of being usable in different contexts. To these were then added what an uncertain Department for Education and Employment still calls '*the other three*' more innovatory skills, which are yet to be made obligatory:

- problem-solving;
- personal skills, improving own learning (autonomy, or self-learning);
- personal skills, working with others.

Most of these had already existed in the vocational science courses run by BTEC in colleges and polytechnics for many years, and were known there as *Common Skills*. This practice in vocational education may be transported into general use in schools and universities, to the immense chagrin of many liberal educationalists. On the whole, however, it is fair to say that more attention is paid to the '*other three*' skills, especially the last two, in the FE colleges which used to teach BTEC, than it is anywhere else.

Improving one's own learning acquires a new meaning in GNVQ courses. Going round colleges where GNVQ science is being taught, it is easy to see that the major concern of the students is almost never the difficulty in understanding the concepts involved, as it might be at A level. It is getting all the assignments, which comprise the course work, completed to time and filed away in their portfolios. This requires steady work to a schedule; what used to be called 'action planning' in BTEC courses.

It's a lot of work but you must organize yourself – time management.

Student

We have to work harder than on A levels. You get assignments from each subject. It's hard to keep up. You have to be organized.

Student

Depending on the particular student, this may either be welcomed as independence, or dreaded as a pressure of work to be done. It is certainly the case that the apparently high number of drop-outs recorded each September in GNVQ is more often due to an incomplete portfolio of coursework than to having left the course. The independence factor is realized through asking students to find out information for themselves, so as to satisfy the criteria for merits or distinction. This, like the rest of the criteria for assessment, is admirably transparent to the students, so that those who are motivated almost exclusively by *achievement* find this very satisfying. This kind of motivation does not come into the category of *deep learning*, as mentioned in an earlier section, but it seems very common among the students to be met on GNVQ.

I prefer it because you can improve your learning by trying to do better in each assignment. You can control it; you know how you are getting along.

Student

The last of the key skills is *working with others*, which employers call *working in teams*, and is not entirely new to either schools or colleges. Practical work, which is emphasized in GNVQ, has always been carried out in groups. The main difference is that, now, even the dreaded *writing-up* can be done together without any sign of disapproval from the tutor. The students notice this, and most of them welcome it. However, it has to be said that there are no signs that the skills involved in the process are ever taught to the students, so they may only be practising old social skills rather than acquiring new vocational ones.

It is harder to find teachers for the post-16 age group who can tackle *communication* in an appropriate way. Far too often, the GNVQ science group is just *sent off to English lessons*, where they are extremely lucky if what they get has any relevance to either science or science-related employment.

The better science teachers recognize that the new skills should be taught within science and have welcomed them. However, it needs to be said that these are only the better teachers. The continual checking to make sure that the students are up to date with coursework and the friction this causes, the need to improvise new materials as the specifications change from year to year, and even the perceived lack of status in having to teach the less able students, takes its toll on the less committed teachers. Fortunately, the best are still very good indeed and are clearly inspired by the challenge.

Teaching GNVQ also makes you a better teacher ... because you have to encompass the different learning styles of the students ... With A level I think almost anyone who knows the material can go ahead and do it – there is a rigidity about it, and you will get some success. But you can't do that at GNVQ. You have to be more reflective about what you are doing – you also need a vision of what you are trying to do because there is no form to the thing ... So, you need to have a bit of vision about what science education is about ...

Tutor

(Teachers who wish to pursue the idea of students with different learning style preferences at GNVQ and A level are advised to read Roy Haywood's (1997) interesting chapter on this theme. It might be fair to conclude from this that, although the preferred learning style does vary markedly from one subject to another at post-16+ level, the variation is less between GNVQ and A level in the same subject.)

Summary

For this selection of research to be of value to schools it needs to be summarized in a way which aids managing the introduction and continuity of GNVQ. For greater clarity we shall abandon any pretensions of literary style and itemize our outcomes in terms of the actions which a school might wish to take.

*1   For cultural and other reasons there is no real parity of esteem with
A level.*

This may be at its most salient when dealing with parents at the start of the course. However, there are at least three counter-vailing points, which can be used.

- Taking GNVQ can prevent youngsters from dropping-out of school and education altogether, and can lead to valuable UCAS points for university entrance.
- The statistically rated 'chances' of getting a good grade at A level, when starting with a D grade at GCSE, are less in science than in any other subject.
- It has been shown to be easier to get UCAS points from GNVQ than from the two A levels in science to which it is supposedly equivalent.

2   It is 'easier' than A level.

- There are no examinations – only tests at the end of modules and coursework. Many students prefer this.

- The knowledge and explanations are given in small chunks and in a particular context. Many students find this easier to learn.
- Students who would do better at A level prefer more abstract conceptual learning.

3   *There is no compulsion to take work experience for GNVQ, although it is recommended for both A level and GNVQ. It is sometimes very difficult to find placements in science, but has considerable value in the context of this and any vocational course.*

- The working-and-learning context helps to develop work values and self-identity.

4   *GNVQ sets out to improve the students' learning skills in connection with transparency of assessment criteria.*

- The students have to develop useful time management skills in order to complete their coursework.
- Students can achieve higher grades by learning more independently and by consulting data sources for themselves.
- Students are encouraged to work in groups.

5   *Staffing choice is very important.*

- Tutors need to welcome the chance to permit and develop student autonomy.
- Problems will arise with students who have difficulties in organizing their time schedule.
- Tutors will need to innovate in order to construct appropriate assignments, since there have been frequent changes in the specifications.
- Finding staff to teach the necessary literary skills, which are relevant to science, is often difficult.

Should the necessary science staff be available, experience suggests that the scope for the exercise of personal vision and teaching action will be much appreciated by them!

References

Barnett, R. (1992) What effects? What outcomes? in R. Barnett (ed.) *Learning to Effect*. Buckingham: The Society for Research into Higher Education and Open University Press.

Biggs, J. (1976) Dimensions of study behaviour: another look at ATI, *British Journal of Educational Psychology*, 46: 68–80.

Burns, J., Clift J. and Duncan, J. (1991) Understanding of understanding: implications for learning and teaching, *British Journal of Educational Psychology*, 61: 276–89.

Callaway, W.J. (1982) *Supervised Work Experience in Educational Programmes: Sandwich Degree Courses in Design, a Case Study.* London: CNAA.

*Fitz-Gibbon, C. (1997) Listening to students, in T. Edwards, C. Fitz-Gibbon, F. Hardman, R. Haywood and N. Meagher (eds) *Separate but Equal?* London: Routledge.

Hayward, G. and Solomon, J. (1993) *A Preliminary Study of Post-16 Education.* Oxford: Oxford University Department of Educational Studies.

*Haywood, R. (1997) Links between learning styles, teaching methods and course requirements in GNVQ and A level, in T. Edwards, C. Fitz-Gibbon, F. Hardman, R. Haywood and N. Meagher (eds) *Separate but Equal?* London: Routledge.

Hyland, T. and Johnson, S. (1998) Of cabbages and key skills: exploding the mythology of core transferable skills in post school education, *Journal of Further and Higher Education,* 22(2): 163–72.

*Jessup, G. (1991) *Outcome: NVQs and the Emerging Model of Education and Training.* London: Falmer Press.

*Lave, J. and Wenger, E. (1991) *Situated Learning: Legitimate Peripheral Participation.* Cambridge: Cambridge University Press.

Prais, S. (1995) *Productivity, Education and Training.* Cambridge: Cambridge University Press.

*Smithers, A. and Robinson, P. (1995) *Post-18 Education, Growth, Change Prospect.* Manchester: CIHE University of Manchester.

Solomon, J. (1994) *Student Case-studies in the Pilot Year of GNVQ Science at Advanced Level.* Oxford: Oxford University Department of Educational Studies (for the Employment Department).

Solomon, J. (1996) *Good Practice in GNVQ Science at Advanced Level: Admissions and Learning in Higher Education.* Oxford: Oxford University Department of Educational Studies (for the Employment Department).

14 Science for citizenship

Jonathan Osborne

'Nobody has died, nobody is sick, nothing has happened' is the caption on an article on genetically modified foods that looms out of the paper. Over the page, 'Stop the crops' articulates the opposite, alarmist view. All of a sudden, supermarkets are full of people scouring labels with a thoroughness that has to be seen to be believed. And the London *Independent*'s recent comment that 'the real challenges for the future are scientific' seem to have a prescience that even the writer could not have foretold. Yet what role, if anything, does science education have to contribute to this and other debates?

Socio-scientific issues, and their accompanying ethical, political and moral concerns, from BSE to global warming, increasingly dominate the media and public and family life. Although science is one of the major achievements of Western civilization, and permeates our culture rather as mica pervades granite, ever since that fearsome mushroom cloud rose over the Nevada desert and Rachel Carson published her book *The Silent Spring*, the pretence that science and scientists are separate from society and its applications has been unsustainable.

Furthermore, the dilemmas posed by modern science have raised concerns about the public understanding of science, first expressed by scientists in the Bodmer Report (1985), and echoed most recently as a need to improve the general 'scientific literacy' of the adult population (*Times* 1998; *Financial Times* 1999). Not to know any science is to be an 'outsider' – an alien to the culture as much as somebody who cannot recognize the cultural referents that are a product of the 'greats' of English literature. But, does the science education that we practise really help to develop the kinds of competencies and knowledge – the scientific literacy – that our future citizens are likely to need? And, should this be one of its purposes?

Such arguments about aims are important because aims matter. A teacher without an aim is like a ship without a rudder, unable to see the route or the strategies that will provide an education which is appropriate to their pupils' current and future needs. So, beginning with a discussion of aims, this chapter explores and summarizes the scholarly arguments and research evidence that have revolved around the purpose and function of science education. And, if science for citizenship or scientific literacy is to be an aim, this chapter then considers what implications for contemporary classroom practice might be.

Aims of science education

Why teach science, and, in particular, why teach science to all pupils? Now that science occupies a privileged position in the school's curriculum in the UK, with up to 20 per cent of curriculum time devoted to its study for pupils between the ages of 14 and 16, this is a question often asked by pupils, albeit in a more simplistic form. While the daily treadmill offers little opportunity to stand and stare and consider such questions, the answers are important in determining the kind of science that is offered to young people, and the emphasis that is given to different aspects of science.

Broadly speaking, there are four arguments for science education, which can be found in the literature (Layton 1973; Milner 1986; Thomas and Durant 1987; Millar 1996). These are called the utilitarian argument, the economic argument, the democratic argument and the cultural argument. In what follows, each of these is described and discussed, and their strengths and weaknesses examined.

The utilitarian argument

This is the view that learners might benefit, in a practical sense, from learning science. That is, scientific knowledge enables them to wire a plug or fix their car; that a scientific training develops a 'scientific attitude of mind', a rational mode of thought, a practical problem-solving ability that is unique to science and essential for improving the individual's ability to cope with everyday life. It is also claimed that science also trains powers of observation, providing an ability to see patterns in the plethora of data that may confront us. Such arguments may well resonate with the reader – they are, after all, the stock-in-trade responses that are part of the culture of science teaching. Sadly, however, they do not stand up to close examination.

First, there is little evidence that scientists are any more or less rational than the rest of humanity. As Millar (1996) argues, 'there is no evidence that physicists have fewer road accidents because they understand Newton's laws of motion, or that they insulate their houses better because they understand the laws of thermodynamics'. Secondly, the irony of living in a technologically advanced society is that we become *less* dependent on scientific knowledge. The increasing sophistication of contemporary artefacts makes their functional failure only remediable by the expert, while simultaneously, their use and operation is simplified to a level that requires only minimal skill. Electrical appliances come with plugs pre-wired, while washing machines, computers, video-recorders, and so on, require little more than intuition for their sensible use. Even in contexts where you might think that scientific knowledge would be useful, such as the regulation of personal diet, recent research on pupils' choice of foods shows that it bears no correlation to their knowledge of what constitutes a healthy diet (Merron and Lock 1998).

Any idea that science trains powers of observation has long been undermined by the recognition that observation is a theory-dependent process (Hanson 1958). Studies of perception reveal that observers tend to pay attention to objects or features with which they are familiar – that is, observers are influenced by the ideas that *they bring to* their looking, and see first what they expect to see. As Driver (1983) showed in her work, pupils often find it difficult to see cheek cells on a microscope slide, or the accepted pattern of filings surrounding a magnet, because they lack a clear concept of what they are looking for. Further support is provided by research which shows that children's performance on observation type tasks is significantly improved (Hainsworth 1956; Bremner 1965) if prior instruction is offered about the kind of structures they expect to observe. The inevitable conclusion to be drawn from such work is that a utilitarian argument for knowledge is open to challenge on a number of fronts. It is, in short, an argument that science teachers would be ill-advised to use with their pupils, and ill-advised to use with headteachers and curriculum managers to justify science's claim for such a large slice of precious curriculum time.

The economic argument

This is the argument that an advanced technological society needs a constant supply of scientists to sustain its economic base and international competitiveness. From this perspective, science is seen as providing a pre-professional training, and acts essentially as a sieve for selecting the chosen few who will enter academic science or follow courses of vocational training. The 'wastage' is justified by the fact that the majority will ultimately benefit from the material gains that the chosen few will provide. The trouble with this argument is the data. The latest statistics for the UK show that, in 1993, there were 32,621 undergraduate entrants to science-based (including engineering, science, technology, and so on) courses. As such, in common with many other countries, they are only a very small percentage of the large number who take science courses to age 16 (400,000+ in the UK) or later. Therefore, to argue that their needs should be the dominant or sole determinant of the aims, and hence the content, of the science curriculum, would be unjust.

Coles (1998) reports that the most systematic and comprehensive analysis of what scientists themselves *do* in the UK was carried out by the Council of Science and Technology Institutes. Their report listed 46 occupations where science was a main part of the job (such as a medical technician), or a critical part of their job (such as a nurse). Some 2.7 million people fell into these categories, a figure which represents only 12 per cent of the UK workforce. A further million people have their work enhanced or aided by a knowledge of science and technology. He estimates that the needs of this group represents, at most, a further 16 per cent of the total UK workforce. Coles's analysis of scientists and their work, their job specifications and other research, summarizes the important components of scientific knowledge and skills needed for employment as:

- general skills;
- knowledge of explanatory concepts;
- scientific skills:
  - application of explanatory concept,
  - concepts of evidence,
  - manipulation of equipment;
- habits of mind:
  - analytical thinking;
- knowledge of the context of scientific work.

Coles's data, collected from interviews with a range of 68 practising scientists, suggest that a knowledge of science is only *one* component among many that are needed for the world of work. Furthermore, his data indicate that the knowledge they *do* need is quite specific to the context in which they are working. In contrast, the scientists in this research stressed the importance of the skills of data analysis and interpretation, and general attributes such as the capacity to work in a team and an ability to communicate fluently, both in the written word and orally – aspects which are currently undervalued by contemporary practice in science education. Baldly stated, even our future scientists would be better prepared by a curriculum that reduced its factual emphasis and covered less, but uncovered more, of what it means to practise science. Coles's findings suggest that the skills developed by opportunities to conduct investigative practical work, such as that required in the UK – the ability to interpret, present and evaluate evidence, the ability to manipulate equipment and an awareness of the scientific approach to problems – are outcomes which are to be valued as much as any knowledge of the 'facts' of science.

The cultural argument

This is the argument that science is one of the great achievements of our culture – the shared heritage that forms the backdrop to the language and discourse that permeate our media, conversations and daily life (Cossons 1993; Millar 1996). In a contemporary context, where science and technology issues increasingly dominate the media (Pellechia 1997), this is a strong argument, succinctly summarized by Cossons.

The distinguishing feature of modern Western societies is science and technology. Science and technology are the most significant determinants in our culture. In order to decode our culture and enrich our participation – this includes protest and rejection – an appreciation/ understanding of science is desirable.

The implication of this argument is that science education should be more of a course in the appreciation of science, developing an understanding not only of what it means to do science, but also of what a hard-fought struggle and great achievement such knowledge represents. Therefore, understanding the culture of science requires some science history, science ethics, science argument and scientific controversy – with more emphasis on the human dimension and less emphasis on science as a body of reified knowledge. In short, this means fewer 'facts' and more of the broad 'explanatory stories' that science offers, together with the development of a better understanding of a range of 'ideas-about-science' (Millar and Osborne 1998).

The democratic argument

Proponents of this view point to the fact that many of the issues facing our society are of a socio-scientific nature. For instance, do we allow cloning of human beings? Should we prevent the sale of British beef? Should we allow electricity to be generated by nuclear power plants? Essentially, the nature of contemporary society has changed from one where science is perceived as a source of solutions to one in which it is *also* seen as a source of problems (Beck 1992). Moreover, as disciplinary knowledge becomes increasingly specialized and fragmented, we become ever-more reliant on expertise. Social systems such as hospitals, railways, and air travel, gain a complexity beyond the comprehension of any individual. Consider, for instance, the number of individuals and systems involved in ensuring the safe flight of one aircraft between London and Paris. In such a context, trust in expert systems and their regulatory bodies plays a large part in our faith that they will function effectively (Giddens 1990).

Unfortunately, the public faith in the expertise of science has been damaged by a number of high-profile failures such as the BSE debacle, Chernobyl, the marketing of unmarked genetically altered foods, and more. If the real challenges of the future are likely to be the moral and political dilemmas set by the expansion of scientific knowledge, a healthy democratic society requires the participation and involvement of all its citizens (or as many as possible), in the resolution of the decisions emerging from the choices that contemporary science will present. This is only likely if individuals have at least a basic understanding of the underlying science, and can engage both critically and reflectively in a participatory debate.

Put simply, scientists, like other members of a democratic society, must be held to account. As a society, we provide large sums to fund and support their research. Should it be directed towards work that promises a material and tangible benefit, for example enhanced food production, a vaccine for malaria, or should we support work which has little obvious benefit, such as the construction of a new, orbiting space station? Most contemporary scholarship would argue that the discussion of such issues would improve if our future citizens held a more critical attitude towards science (Irwin 1995; Fuller 1997; Norris 1997), one which, while acknowledging its strengths, also recognized its limitations and ideological commitments. However, it is difficult to see how this can be done by a science education which offers no chance to develop an understanding of how scientists work, how they decide that any piece of research is 'good' science, and which, in contrast to the controversy and uncertainty that surrounds much contemporary scientific research, offers a picture of science as a body of knowledge which is 'unequivocal, uncontested and unquestioned' (Claxton 1997).

Education for citizenship

What kind of education would help our future citizens to choose between their genetically modified tomato puree or its natural equivalent? Gee (1996) argues that becoming 'literate' means becoming knowledgeable and familiar with the discourse of the discipline. That is, the 'words, actions, values and beliefs of scientists', their common goals and activities, and how they act, talk, and communicate. Such knowledge has to be acquired through exposure to the practices of scientists, and explicitly taught so that children can become critically reflective. Rather as learning a language requires children to develop a knowledge of the form, grammar and vocabulary, so becoming scientifically literate would require a knowledge of science's broad themes, the reasons for belief, at least some of its content and, in particular, its uses and abuses.

The research evidence on 'scientific literacy', however, leads to a questioning of the achievements of science education, for attempts to measure civic scientific literacy paint a depressing picture of the average person's knowledge of science. Since 1987, a number of well-funded surveys have been conducted in the UK (Durant *et al.* 1989), Europe and the United States (Miller *et al.* 1997). These surveys have been conducted using a mix of closed questions using true–false quizzes containing items such as 'Is it true that: "lasers work by focusing sound waves?", "all radioactivity is man made?", "antibiotics kill viruses as well as bacteria?"', and open questions. The open questions asked respondents to tell the interviewer in their own words, for instance, 'What is DNA?' In addition, questions were included that assessed the public's understanding of the procedures of science.

Combining these data with those for their understanding of process, Millar *et al.* (1997) arrived at a view of the numbers who could be deemed to be civically scientifically literate, or at least partially so (Table 14.1).

If we accept Miller's (1998) assertion that the only individuals capable of acquiring and comprehending information about science and technology policy are the small percentage who possess civic scientific literacy – that is,

Table 14.1 Estimated percentage of adults qualifying as civically scientifically literate by country/area

	Literate (%)	Partially literate (%)	Not literate(%)	Number in sample
United States	12	25	63	2,006
European Union	5	22	73	12,147
Britain	10	26	64	1,000
Germany	4	24	72	1,000

Source: adapted from Miller (1998)

somebody who understands some of the procedures of science (its grammar) as well as its content (vocabulary) – then these results beg the question as to why the systematic exposure to formal science education within schools is failing to generate a more scientifically literate populace. The only solace or comfort to be found in the studies is that the most significant determinant of the public's level of knowledge of science – their civic scientific literacy – is the number of years that they have spent studying science full-time (Lucas 1987; Miller 1997).

However, it should be said that this research is subject to a number of criticisms. While such data portray the public as deficient and lacking in scientific knowledge (Wynne 1991; Ziman 1991), a range of studies carried out in a variety of contexts – with sheepfarmers in Cumbria coping with the aftermath of Chernobyl (Wynne 1996); parents of Down's syndrome babies; individuals living near a chemical plant; and electricians at Sellafield (Layton *et al.* 1993) – all demonstrate that the public can engage with scientific expertise in a manner which is both locally and contextually situated, and can acquire new scientific knowledge on a need-to-know basis. Just as one's memory of a foreign language fades into the dim and distant past without use, so do the 'facts' of science unless there is a regular need to use such information. Hence, the results obtained by such surveys of the public knowledge of science are unexceptional, and similar data could be obtained if we were to survey the public understanding of English literature.

A succinct summary of this research is provided by Jenkins (1998), who argues that it shows that the interest of citizens in science is differentiated by the science, social group and gender. For most of us, interest in science is related to decision-making or action, and that when necessary, we choose a level of knowledge adequate for the task-in-hand and learn what is essential. As Jenkins concludes (1998; 12), 'the "non-expert" citizen turns out to be rather complex in his or her dealings with science' and those interactions 'cannot be explained simply in terms of ignorance'.

For science education, such findings have an important message suggesting that the over-emphasis on content, coupled with the negative attitudes it engenders (see Chapter 7) is a wasted endeavour. If the public's retention of the disciplinary content of science is so situationally specific, would it not be better to spend time developing an understanding of what science is, how it is done, its broad areas of study and the major ideas that it has contributed to our culture, rather than attempting to construct an understanding 'brick' by 'brick' or 'fact' by 'fact'? After all, the latter process would not appear to develop a structure of any permanence. It is true that one cannot have a knowledge of science without acquiring some of its major conceptual ideas, or without understanding some of the methods it uses to justify its claims. But, as Gee (1996) argues, teaching for acquisition alone leads to successful but 'colonized' students, who have no knowledge about their own discipline, such as its history or its evidential base, leaving them bereft of many of the faculties necessary to engage critically with the assertions and opinions of scientists, for instance, that eating genetically modified foods is absolutely safe. Too much emphasis on content then leads to 'too little analytic and reflective awareness and limits the capacity for certain sorts of critical reading and reflection'. History teachers, in contrast, have made such a transition. They now see their subject as one in which content is subsidiary to developing the skills of historical analysis needed to make sense of contemporary life and resolve uncertainty (Donnelly 1999). Is it not time for science education to make a similar paradigmatic shift?

Implications for science teaching

Teaching *about* science would require that more emphasis is given to developing: an understanding of the methods and processes of science; an awareness of the context and interests of scientists, their social practices; and a capacity to analyse, or at least consider, risks and benefits (AAAS 1989, 1993; Millar and Osborne 1998). As Ziman (1994) argues, merely teaching about the applications of science is insufficient, for such approaches simply take a quick stride from science to technology, but usually fail to go on and consider the societal implications, thus perpetuating the notion that science offers a technical fix for all our problems. However, contemporary science raises issues whose solution requires careful consideration of ethical and moral values, for example: should we grow genetically modified organisms; should we restrict car use, and so on? While the market ideology that dominates the production of school science curricula argues that the consideration of such issues is irrelevant, the failure to incorporate such elements leads to a widening gulf between science-as-it-is-taught and science-in-the-media.

The Science–Technology–Society (STS) movement sought, in contrast, to situate learning about science in a social context, arguing that the presentation of science as an academic, value-free subject was seriously out of date. This led to a series of courses and materials, such as SISCON (Solomon 1983), ChemCom: Chemistry in the Community (American Chemical Society 1988) and SATIS (ASE 1986), which provide a range of resources for introducing such issues; they are too often under-used, and are an important first port of call for teachers interested in situating the science they teach in a broader context.

More guidance for developing the skills and competencies for citizenship is dependent on the brief lights shone by specific pieces of research in a range of areas, and the substantial body of work that has been undertaken in the history, philosophy and sociology of the subject in the past 30 years. From such work, several areas stand out for attention in school science education.

Argument in science

When scientific claims are made, theories are often challenged, and progress is made through dispute and conflict (Kuhn 1962; Taylor 1996; Fuller 1997). Assessing alternatives, weighing evidence, interpreting texts, evaluating the potential viability of scientific claims are all essential components in evaluating scientific arguments (Latour and Woolgar 1986). Science-in-the-making is also always characterized by a number of uncertainties: empirical uncertainty due to lack of evidence; pragmatic uncertainty due to a lack of resources to investigate the problem; and theoretical uncertainty due to a lack of a clear theory of what is causing the events of interest. Increasingly, arguments between scientists extend into the public domain through journals, conferences and the wider media, and it is only through such processes – checking claims and public criticism – that 'quality control' in science is maintained.

Yet, as currently practised, science education uses evidence to persuade pupils that the *singular* account offered by the teacher is self-evident and 'true'. There is little attempt to develop an understanding of the logic and reasoning that is used to argue for, or against, a scientific hypothesis (Giere 1991). This contrast, and gulf, between science-as-practised and science-as-taught can only be resolved if pupils are occasionally given the opportunities to study more than one interpretation of a set of data and critically examine the arguments for both cases (Siegel 1989; Geisler 1994; Monk and Osborne 1997). Research on individuals' abilities to argue from evidence to conclusions (Kuhn 1993; Kuhn *et al.* 1997) suggests that the majority of individuals display a naivety in their argumentative skills. Kuhn found that individuals display a range of errors in reasoning such as 'false inclusion' (essentially seeing correlations between two variables as being causal); the failure to use exclusion (that is, controls – a method essential to scientific reasoning as it allows the elimination of extraneous factors from consideration); the domination of affirmation over negation (looking for evidence to confirm an idea when scientific ideas survive because they are *not* disproved); and a tendency to dismiss factors as irrelevant (thus eliminating the potential for disconfirming evidence). Somewhat dishearteningly, Kuhn found that schooling made no difference after the end of junior high school, a finding which suggests that there is too little attention paid to the practice of reasoning and argument in high schools.

The implication is that, rather than presenting science as a series of successful discoveries, young people should occasionally be offered the opportunity to study aspects of science-in-the-making so that they can begin to understand why scientists might disagree, and why so much uncertainty surrounds scientific work at the boundaries of our knowledge. One approach would be to undertake more detailed case studies of scientific discoveries such as those suggested by Matthews (1994), Solomon (1991, 1992) and Osborne (1998). These show that, whenever a new explanation of a phenomenon is offered, there are always at least two, if not more, competing theoretical interpretations put forward. Resolution often takes many years – as it did with Galileo's arguments for the heliocentric theory of the solar system; Torricelli's assertion that there was a vacuum at the top of the barometer when there were good logical arguments why 'nature abhorred a vacuum'; or Wegener's almost lunatic (at the time) assertion that the continents had once been one and drifted apart. The other advantage of such case studies is that they re-introduce into science the aspect that often seems to be missing for so many students – people. Thus stories can be told about Joule on his honeymoon, Marie Curie and her lover Langévin, Pasteur's deceit with his anthrax vaccine, Rosalind Franklin circulating scornful, black-edged cards announcing the death of DNA helix and more – all of which add an essential extra human dimension to the practice of science, which has been systematically erased from standard texts.

Developing an understanding of evidence

Another task is to develop a more substantive grounding in what constitutes evidence in science – what Gott *et al.* (1997) term 'concepts of evidence'. This requires that much more time be given to exercises which require the quality of data to be assessed: how accurate are they, how much error was present in their measurement, and how much can they be trusted? (Pritchard 1989; Gott *et al.* 1997). In science lessons, a regular feature should be exercises in transforming data from one form to another, from tables to graphs and vice versa, so that students can develop fluency in a skill which is not only essential to evaluating scientific findings but which also has value far beyond the boundaries of science wherever bodies of data are used to support arguments.

It is also somewhat strange that much of science education as practised shows an obsessive concern with the methods of the physical sciences. Science is presented as a form of empirical enquiry, based solely on a hypothetico-deductive model of investigation. In contrast, much research reported in the media is based on epidemiological or correlational studies (Bencze 1996), with the use of controls, and blind or double-blind testing – concepts which are rarely even mentioned, let alone modelled in the science classroom. Yet, simple exercises tabulating hair colour against eye colour, or hours spent watching television against hours spent doing homework, open a window into one of the principal methods of science.

Contemporary science

Some illumination about where science educators might concentrate more of their efforts comes from the growing body of scholarship emerging from the study of science communication, particularly that in the popular press. After all, as Nelkin (1995: 2) argues, 'for most people the reality of science is what they read in the press. They understand science less through direct experience or past education than through the filter of journalistic language and imagery.' One 1997 study, which examined the trends in science coverage in three major US daily newspapers over the period 1966–90, found that the disciplinary divisions were medicine, health, nutrition and fitness (73 per cent), technology (5 per cent) and natural/physical sciences (22 per cent) (Pellechia 1997). The implication, therefore, is that biological science and its methods is *the* important science to address if we wish to provide a knowledge that will be both valued and valuable.

Providing an opportunity to read and discuss contemporary reports about science offers another means of extending pupils' ability to understand and interpret science (Wellington 1991). Researchers in the area of science communication (Perlman 1974; Hinkle and Elliot 1989; Evans *et al.* 1990) have argued that the following components are important factors in determining the importance and significance of any reported science story:

- the location and length of coverage, as these give a measure of its importance;
- the source of the original research, as papers published at conferences have less prestige than, for instance, those published in *Nature*;
- the identification of the researcher(s) by name and their professional status, for example Dr, Professor, as this enables some discrimination about the level of significance to give to the report;
- the institutional affiliation of the researcher(s), for example university, government, industry, as such information enables us to judge, at least in part, whether the interpretation of the findings might be coloured by the allegiances, commitments and values of the researcher;
- comments from the researcher(s) who conducted the study(ies), which indicates that the report is at least attempting to offer their own interpretation of the findings;
- comments from other scientists: one characteristic of science is that it is a process of organized scepticism. The natural procedure of science is to examine all findings with a view to disbelief. Quotes, or comments in support, suggest that the findings have at least convinced some of the peer community of the value of the findings, whereas comments expressing disbelief warn that the findings are contested;
- contextual factors: media reports have a tendency to characterize scientific research in terms of 'breakthroughs', resulting in over-sensationalization of what might be quite tentative findings. Therefore, one important characteristic is information that informs the reader whether these findings accord with, or deviate from, previous findings.

Developing the ability to read science in a critical, 'educated' manner requires opportunities to explore some of these issues in the science classroom. Again, most of the background knowledge underpinning this set of evaluative criteria is not a knowledge of science itself, but a knowledge of how science is practised, that is, science in its social context; knowledge that will only be developed by the occasional opportunity to read, discuss and explore contemporary science. Evidence that formal science education currently leaves pupils ill-prepared to make such judgements emerges from the work of Norris and Phillips (1994), which showed that, of a sample of 91 able, grade 12 Canadian science students, less than 50 per cent could identify causal statements, fewer than 10 per cent recognized justifications, and 42 per cent confused statements of evidence and conclusions in reading media accounts of science.

Exploring the ethics and values of science

Finally, science does not exist in a vacuum; its practice raises important moral and ethical issues for society, which pupils want to consider. Research into systematic attempts to consider the ethical aspects of science and science-related issues has been conducted by Fullick and Ratcliffe (1996) and Solomon *et al.* (1992). Such work has been a feature of curriculum initiatives such as SATIS (ASE 1986), the Salter's Science Course, and the Canadian course *Logical Reasoning in Science and Technology* (Aikenhead 1991). The approach is usually based around group discussions of socio-scientific issues, from the local and specific, such as the type of materials to use for window frames, to the global, such as what can be done to solve the world food problem. All research and findings on such activities reinforce the view that a clear structure must be provided, both for the conduct of the discussion (Dillon 1994), and for the identification of options and evaluating their relative merits (Ratcliffe 1997). Such exercises provide opportunities to practise and develop an 'an armoury of essential skills: listening, arguing, making a case, and accepting the greater wisdom or force of an alternative view', which are the foundations of responsible citizenship in a participatory democracy (Advisory Group for Education for Citizenship 1998: 61). The implications and consideration of genetic engineering, the re-processing of nuclear fuel, ozone depletion and other such issues are naturally located within the science classroom. Moreover, research shows that their immediate and contemporary relevance is welcomed by pupils (Solomon 1992). Research is now also emerging that would suggest that engaging in such activities does begin the process of enhancing the pupils' skills of argumentation, while simultaneously providing interest and relevance (Zohar and Namet 1998; Driver *et al.* in press).

Conclusions

This chapter has attempted to explore the arguments for science education, and, in particular, for a science education that is a preparation for citizenship rather than life as a professional scientist. Studies of science, as used in the media and everyday life, suggest that 'citizens' science' requires less emphasis on the 'facts' of science and a broader knowledge of how science works. While work in this field is limited, it rests on a premise that 'to know science' is a statement that one knows not only *what* a phenomenon is, but also *how* it relates to other events, *why* it is important and *how* this particular view of the world came to be. Any science education which offers only aspects of science in isolation – attempting to divorce science, technology and the social context of its production – will fail to provide an adequate education for the future citizen. For current practice is rather like introducing a young child to jigsaws by giving them bits of a 1000 piece puzzle, and then only the pieces from the left-hand side, hoping that they have enough to get the whole picture. Surely, providing the simplified and complete 100-piece version might develop more confidence, trust and intellectual independence and would permit a more rational, balanced and informed decision about whether to buy that tube of genetically modified tomato puree?

References

Advisory Group for Education for Citizenship (1998) *Education, Citizenship and the Teaching Democracy. Final Report of the Advisory Group on Citizenship* (98/155). London: Qualifications and Curriculum Authority.

Aikenhead, G.S. (1991) *Logical Reasoning in Science and Technology*. Toronto, Ontario: John Wiley of Canada.

American Association for the Advancement of Science (AAAS) (1989) *Project 2061: Science for All Americans*. Washington, DC: American Association for the Advancement of Science (AAAS).

American Association for the Advancement of Science (AAAS) (1993) *Benchmarks for Scientific Literacy*. Washington, DC: American Association for the Advancement of Science (AAAS).

American Chemical Society (1988) *ChemCom: Chemistry in the Community*. Dubuque, IA: Kendall/Hunt.

Association for Science Education (ASE) (1986) *Science and Technology in Society (SATIS)*. Hatfield: Association for Science Education.

*Beck, U. (1992) *Risk Society: Towards a New Modernity*. London: Sage.

*Bencze, J.L. (1996) Correlational studies in school science: breaking the science-experiment-certainty connection, *School Science Review*, 78(282): 95–101.

*Bodmer, W.F. (1985) *The Public Understanding of Science*. London: The Royal Society.

Bremner, J. (1965) Observation of microscopic material by 11–12 year old pupils, *School Science Review*, 46: 385–94.

Claxton, G. (1997) A 2020 vision of education, in R. Levinson and J. Thomas (eds) *Science Today*. London: Routledge.

Coles, M. (1998) The nature of scientific work: a study of how science is used in work settings and the implications for education and training programmes. Unpublished PhD, Institute of Education, London.

Cossons, N. (1993) Let us take science into our culture, *Interdisciplinary Science Reviews*, 18(4): 337–42.

*Dillon, J.T. (1994) *Using Discussion in Classrooms*. Buckingham: Open University Press.

Donnelly, J. (1999) Interpreting differences: the educational aims of teachers of science and history, and their implications, *Journal of Curriculum Studies*, 31(1): 17–41.

Driver, R. (1983) *The Pupil as Scientist?* Milton Keynes: Open University Press.

Driver, R., Newton, P. and Osborne, J. (in press) Establishing the norms of scientific argumentation in classrooms, *Science Education*.

Durant, J.R., Evans, G.A. and Thomas, G.P. (1989) The public understanding of science, *Nature*, 340: 11–14.

Evans, W.A., Krippendorf, M., Yoon, J.H., Posluszny, P. and Thomas, S. (1990) Science in the prestige and national tabloid presses, *Social Science Quarterly*, 71: 105–17.

Financial Times Editorial (1999) The perversion of science, *Financial Times*, 20 February.

*Fuller, S. (1997) *Science*. Buckingham: Open University Press.

Fullick, P. and Ratcliffe, M. (1996) *Teaching Ethical Aspects of Science*. Southampton: Bassett Press.

Gee, J. (1996) *Social Linguistics and Literacies*, 2nd edn. London: Taylor and Francis.

Geisler, C. (1994) *Academic Literacy and the Nature of Expertise: Reading, Writing and Knowing in Academic Philosophy*. Hillsdale, NJ: Lawrence Erlbaum.

*Giddens, A. (1990) *The Consequences of Modernity*. Cambridge: Polity Press.

Giere, R. (1991) *Understanding Scientific Reasoning*, 3rd edn. Fort Worth, TX: Holt, Rinehart and Winston.

Gott, R., Foulds, K., Johnson, P., Roberts, R. and Jones, M. (1997) *Science Investigations*. London: Collins Educational.

Hainsworth, M.D. (1956) The effect of previous knowledge on observation, *School Science Review*, 37: 234–42.

Hanson, N.R. (1958) *Patterns of Discovery*. Cambridge: Cambridge University Press.

Hinkle, G. and Elliot, W.R. (1989) Science coverage in three newspapers and three supermarket tabloids, *Journalism Quarterly*, 66: 353–8.

*Irwin, A. (1995) *Citizen Science*. London: Routledge.

Jenkins, E. (1998) *Scientific and Technological Literacy for Citizenship: What Can We Learn from the Research and other Evidence?* Available from http://www.leeds.ac.uk/educol/documents/000000447.doc.

*Kuhn, D. (1993) Science as argument: implications for teaching and learning scientific thinking, *Science Education*, 77(3): 319–37.

Kuhn, D., Shaw, V. and Felton, M. (1997) Effects of dyadic interaction on argumentative reasoning, *Cognition and Instruction*, 15(3): 287–315.

Kuhn, T.E. (1962) *The Structure of Scientific Revolutions*. Chicago: University of Chicago Press.

*Latour, B. and Woolgar, S. (1986) *Laboratory Life: The Construction of Scientific Facts*, 2nd edn. Princeton, NJ: Princeton University Press.

Layton, D. (1973) *Science for the People: The Origins of the School Science Curriculum in England*. London: Allen and Unwin.

Layton, D., Jenkins, E.W., McGill, S. and Davey, A. (1993) *Inarticulate Science? Perspectives on the Public Understanding of Science*. Driffield: Nafferton, Studies in Education.

Lucas, A.M. (1987) Public knowledge of radiation, *The Biologist*, 34(3): 125–9.

Matthews, M.R. (1994) *Science Teaching: The Role of History and Philosophy of Science*. New York: Routledge.

Merron, S. and Lock, R. (1998) Does knowledge about a balanced diet influence eating behaviour? *School Science Review*, 80(290): 43–8.

*Millar, R. (1996) Towards a science curriculum for public understanding, *School Science Review*, 77(280): 7–18.

Millar, R. (1998) Rhetoric and reality: what practical work in science education is *really for*, in J. Wellington (ed.) *Practical Work in School Science: Which Way Now?* London: Routledge.

*Millar, R. and Osborne, J.F. (eds) (1998) *Beyond 2000: Science Education for the Future*. London: King's College London.

Miller, J.D. (1997) Civic scientific literacy in the United States: a developmental analysis from middle-school through adulthood, in W. Gräber and C. Bolte (eds) *Scientific Literacy*. Kiel: IPN.

Miller, J.D. (1998) The measurement of civic scientific literacy, *Public Understanding of Science*, 7: 203–23.

Miller, J.D., Pardo, R. and Niwa, F. (1997) *Public Perceptions of Science and Technology: A Comparative Study of the European Union, the United States, Japan, and Canada*. Madrid: BBV Foundation.

Milner, B. (1986) Why teach science and why to all? in J. Nellist and B. Nicholl (eds) *The ASE Science Teacher's Handbook*. London: Hutchinson.

Monk, M. and Osborne, J. (1997) Placing the history and philosophy of science on the curriculum: a model for the development of pedagogy, *Science Education*, 81(4): 405–24.

Nelkin, D. (1995) *Selling Science: How the Press Covers Science and Technology*. New York: Freeman.

Norris, S. (1997) Intellectual independence for nonscientists and other content-transcendent goals of science education, *Science Education*, 81(2): 239–58.

Norris, S. and Phillips, L. (1994) Interpreting pragmatic meaning when reading popular reports of science, *Journal of Research in Science Teaching*, 31(9): 947–67.

Osborne, J.F. (1998) Learning and teaching about the nature of science, in M. Ratcliffe (ed.) *ASE Guide to Secondary Science Education*. Cheltenham: Stanley Thornes.

Pellechia, M. (1997) Trends in science coverage: a content analysis of three US newspapers, *Public Understanding of Science*, 6: 49–68.

Perlman, D. (1974) Science and the mass media, *Daedalus*, 103: 207–22.

*Pritchard, I. (1989) *Data Handling Skills*. Cambridge: Cambridge University Press.

Ratcliffe, M. (1997) Pupil decision-making about socio-scientific issues within the science curriculum, *International Journal of Science Education*, 19(2): 167–82.

*Siegel, H. (1989) The rationality of science, critical thinking and science education, *Synthese*, 80(1): 9–42.

Solomon, J. (1983) *Science in a Social Context (Siscon)-in-Schools*. Oxford: Basil Blackwell.

Solomon, J. (1991) *Exploring the Nature of Science: Key Stage 3*. Glasgow: Blackie.

Solomon, J. (1992) The classroom discussion of science-based social issues presented on television: knowledge, attitudes and values, *International Journal of Science Education*, 14(4): 431–44.

*Solomon, J., Duveen, J. and Scott, L. (1992) *Exploring the Nature of Science: Key Stage 4*. Hatfield: Association for Science Education.

Taylor, C. (1996) *Defining Science: A Rhetoric of Demarcation*. Madison, WI: The University of Wisconsin Press.

Thomas, G. and Durant, J. (1987) Why should we promote the public understanding of science? in M. Shortland (ed.) *Scientific Literacy Papers*. Oxford: Oxford Department of External Studies.

Times Editorial (1998) Hard science: experiments in the laboratory of the future, *The Times*, 11 September.

*Wellington, J. (1991) Newspaper science, school science: friends or enemies? *International Journal of Science Education*, 13(4): 363–72.

Wynne, B. (1991) Knowledges in context, *Science, Technology and Human Values*, 16: 111–21.

Wynne, B. (1996) May the sheep safely graze? in S. Lash, B. Szerszynski and B. Wynne (eds) *Risk, Environment and Modernity*. London: Sage Publications.

Ziman, J. (1991) Public understanding of science, *Science, Technology and Human Values*, 16(1): 99–105.

Ziman, J. (1994) The rationale of STS education is in the approach, in J. Solomon and G. Aikenhead (eds) *STS Education: International Perspectives on Reform*. New York: Teachers College Press.

Zohar, A. and Namet, F. (1998) Fostering argumentation skills through bio-ethical dilemmas in genetics. Conference paper, University of Göteborg, Sweden, the Second Conference of European Researchers in Didactics of Biology.

Index